MEDICINAL PLANTS
OF THE DESERT AND CANYON WEST

The American Southwest is a rich source of medicinal plants and knowledge about their uses. Indian, Mexican, and European peoples live together in this region, and all have their traditions of herbal remedies. Michael Moore has collected much of this information, organized it for practical use, and added his own observations in the present book. *Medicinal Plants of the Desert and Canyon West* is a valuable reference and guide to the herbal remedies of this distinctive ecosystem.

As a botanist-physician, I believe that intelligent use of healing plants can make people more self-reliant in matters of health. In my own practice I use herbal preparations far more than pharmaceutical drugs. I have never yet produced a significant adverse reaction from giving patients plants. More important, I have seen a great deal of benefit from their use as part of a comprehensive program of treatment.

Useful information on herbal medicine is hard to come by except from people who actually do it. Books are often either too technical—unedited for the practitioner; or too general—lacking sufficient data about composition and specific actions. Michael Moore's writings strike a balance between these extremes. Coming from a practicing herbalist, they tell the reader just how to find, prepare, and use the species described, and they also summarize what is known of the plants' chemistry and pharmacology.

I have often consulted the author's previous work, *Medicinal Plants of the Mountain West,* and I expect to be a frequent reader of this new volume. It is a welcome addition to the literature on the Southwest and on natural medicine.

Andrew Weil, M.D.,
Division of Social Perspectives in Medicine,
University of Arizona College of Medicine,
Tucson, Arizona

MEDICINAL PLANTS
OF THE DESERT AND CANYON WEST

A guide to identifying, preparing, and using traditional medicinal plants found in the deserts and canyons of the West and Southwest.

By
Michael
Moore

Illustrations by
Mimi Kamp and
Nora Ryerson

MUSEUM OF NEW MEXICO PRESS

This book is dedicated to my folks, Mary and Wilbur, and to Susan, my friend. I guess I *will* become a herbalist when I grow up.

This publication was made possible through support of the State of New Mexico's Office of Cultural Affairs' Publications Revolving Fund. The Museum of New Mexico Press is a unit of the Museum of New Mexico, a division of the State Office of Cultural Affairs.

Library of Congress Cataloging-in-Publication Data

Moore, Michael, 1941–
    Medicinal plants of the desert and canyon West.

    Bibliography: p.
    Includes index.
    1. Medicinal plants — West (U.S.) 2. Medicinal Plants — West (U.S.) — Identification. 2. Materia medica. Vegetable — West (U.S.) I. Title [DNLM: 1. Desert Climate. 2. Plants, Medicinal — United States. QV 770 AA1 M8ma]
QK99.U6M65 1987   615'321'0978   87-21954
ISBN 0-89013-181-3
ISBN 0-80913-182-1 (pbk.)

The publisher assumes no responsibility for the effects of the techniques used in this book.

Interior design by Daniel Martinez
Cover design by Bruce Taylor Hamilton
Cover photograph by Charles Mann
Printed in the United States of America

10 9 8

Museum of New Mexico Press
P.O. Box 2087
Santa Fe, New Mexico 87504

CONTENTS

ACKNOWLEDGEMENTS

A number of people have helped me here and there, in the field, in the back rooms of *botanicas* and herb stores on both sides of the border—pharmists, physicians, midwives, nurses, *curanderas*, natural products chemists, academics of all persuasions (who bemusedly presume me to represent some *other* discipline than theirs). Bits and pieces, information and tricks of the trades shared between fellow herb freaks. Thanks.

Specifically, I had lots of help and advice and interdisciplinary guidance from Steven Dentali, Charles Mason and Andrew Weil of the University of Arizona, the Crosswhites of Boyce-Thompson, Kevin and Marla Kopriva; my fellow greybeards, Michael Gregory and Peter Bigfoot; my longtime Flagstaff connection, Phyllis Hogan and Sam Boone; Greg Cajete of IAIA, and The Keeper of the Record for contemporary American herbalism, Steven Foster. And thanks to my long-suffering and non-violent editor, Mary Wachs, who kindly only ground her teeth when I wasn't in her presence. Also, thanks to Nora Ryerson, who nearly washed into lakes to get some of these illustrations, and Mimi of Bisbee Junction who also got me locations and contacts. . .and drawings.

FOREWORD

My first introduction to Michael Moore's writing came a decade ago. At the time I served as botanical editor for *Well-Being* magazine. Publisher David Copperfield had just received galleys of Michael's *Medicinal Plants of the Mountain West*. He had been invited to prepare a foreword to the book. David handed me the manuscript for comments. I agreed with his assessment that it was the best herbal to come along in years. That book has since served as the primary reference on medicinal plants of the Western U.S.

Finally, the long awaited companion, *Medicinal Plants of the Desert and Canyon West*, has arrived. While many books on herbs are borne of what Dr. James Duke of USDA's Germplasm Services Laboratory calls "bibliographical echo," or the "Gerard said this, Culpepper said that" syndrome, Moore's books are products of unique and broad practical knowledge. Knowledge provides wisdom rather than the hollow words of rote information. Michael is a combination of gatherer, pharmacist, clinician, teacher, and writer. When he writes about identifying or harvesting a particular plant, he draws from his own experience. When he writes of medicinal use, he does so as an experienced clinician. This combines with an ability to articulate detail in a clear, concise, readable, even humorous style. Among his peers, Michael is known as an herbalist's herbalist. He started practicing his craft more than twenty years ago—before "herbs" became posh. Mark Blumenthal, executive director of the American Botanical Council and editor of *Herbalgram*, calls Moore the "Godfather of American herbalists."

All comments on the author aside, what makes this book a major contribution to the world's medicinal plant literature is the biomes that harbor the medicinal plants of which he writes. During the heyday of conventional medical interest in American botanicals from the 1790s through the 1920s, the vast majority of writing focused on species of the eastern United States, primarily the eastern deciduous forest, adventive European and Asian species, and cosmopolitan weeds. Besides obscure, seldom read, specialized ethnobotanies, little has ever been written about the medicinal plants of inter-mountain deserts, shrub steppes, woodlands, chaparral, and the warm deserts of the Southwest.

In order to survive the stressful xeric environment, intensely hot in summer, harshly cold by winter, plants of these biomes have undergone rapid evolutionary specialization. Like the ability of some highly competitive weeds to dominate habitat through production of secondary metabolites which thwart the growth of other species, many plants of the desert have become chemical factories, producing compounds that help them to survive. It is these secondary metabolites that provide medicinal activity for humans.

Most American scientists working on medicinal plants focus on the rich diversity of threatened tropical rain forests. This has left a void, where unique biomes of North America, like those of the desert Southwest, have been largely neglected. The deserts, too, are rich in vegetation: highly specialized plants producing compounds to repel invaders, attack pathogens, or revitalize and strengthen the organism itself.

We err in protecting endangered species without protecting their habitat, for the habitat itself is a living organism. Nowhere is this fact more dramatic than in the desert and canyon enclaves of the Southwest. Since the habitat will not comfortably support large-scale urbanization, vast expanses have remained essentially intact, though simply cutting a road through desert "wasteland" impacts its biota forever. The botanical pharmacopeia, dependent upon a plant's response to and ability to survive in a highly complex and competitive environment, can evolve only if we understand the need to maintain (let alone) the habitat in which it occurs. Moore exposes the botanical wealth of the desert, furthering the need for their protection.

We are at the threshold of a new era in which botanical products will become more widely known and utilized, as they are in Japan and West Germany. Today, to get the best, latest information on American medicinal plants, one must turn to German scientific periodicals. Biotechnology offers the opportunity to begin to reduce costs for investigation of medicinal plants, while exploring the value of interactive whole plant extracts, rather than isolated, synthesized chemical models, the passing mainstay of drugs from natural products. Pharmacologists verge on developing models for evaluating the pharmacokinetics of whole plant extractives. The first step to developing, enhancing, and utilizing the medicinal value of botanicals is to inventory the plants and recognize the need to preserve habitat diversity.

Michael Moore surveys the desert's medicinal plants, revealing previously unknown resources. He does so in a way that perpetuates the need to protect the habitat and the individual plants themselves. The book sets a precedent for appropriate exploration, development, and protection of the medicinal plant resources of the desert Southwest and points beyond.

Steven Foster,
Research Director,
Ozark Beneficial Plant Project

INTRODUCTION

It has been eight years since I wrote *Medicinal Plants of the Mountain West* for the Museum of New Mexico Press. A lot has changed in the field of botanical medicine in those intervening years . . . and not much has changed, either. There are more (and better) books that have been published than there used to be. There are more (and better) people using botanical medicines than there used to be. Yet, still, the dichotomy between regular school medicine and wholistic medicine grows, the stridency *against* alternative medicine grows in the medical journals, the stridency *against* standard practice medicine grows in the wholistic community. Sophistication has increased . . . so has the finger-pointing on both sides. My response, growingly, is to flip my middle fingers (lovingly) in both directions . . . and heed my own council of the middle way.

One nice change I encountered while preparing this book is the growing body of research material and monographs that are being made available regarding plants, constituents, pharmacology, and the like. As always, most of this comes from journals, sources, and research outside the United States. In the U.S., the work being done is mostly "pure" science, such as chemotaxonomy, natural products chemistry, and other important but largely academic processes. "Applied" research, where the results reported relate to chemistry/pharmacy/medicine/traditional use, is rarely done in this country. Fortunately, the money needed to research and publish articles about the herbal medicines used in India, China, Mexico, or Nigeria is not prohibitive; the Ph.D. candidates are anxious in those countries; the medical researchers are curious; the overview is accomplished; the papers are published; the scientific communities in those countries (and others) understand the need to have their traditions understood, validities validated, perhaps even incorporated into some of their medical practice. In the United States? . . . forget it. We are able to develop and finance BIG medicines; we have no method of developing and financing little medicines (like herbs).

In this country we are embroiled in a grim, desperate, multi-billion-dollar mud-wrestling match between the public sector (the Food and Drug Administration) and the private sector (the pharmaceutical/medical/hospital industry). The initial cost of developing and marketing a new drug (over $50,000,000) makes it financially imperative that no less than a million people a day take that new drug in order to justify and finance its initial research and development expenses. Remember, medicine is our largest industry, bigger than the Pentagon; it costs us 10 percent of our Gross National Product. And is it ever controlled! Restrictions, insurance, forms, reviews, constraints, standard practice red-lining through communities, and the working doc having virtually no freedom of choice, just the recommended (and insured) procedures. With only this one Brobdignagian game in town, everything else slips between the cracks of the two mud wrestlers (admittedly a poor simile).

In Great Britain you can go get some sensible conservative medical treatment from government-run clinics. Or if you want to go medically high on the hog (and can afford it), you can go to a private London hospital and run the high-tech gamut. If you can afford to pay the tariff, you can also get homeopathy or acupuncture or drugless therapy, or treatment by a medical herbalist, knowing that the practitioners are licensed and competent at what they do. Because these latter have been granted an accepted (if marginal) professional standing and a (if begrudgingly) legal status, they are in an enviable position of being able to refer you to medical or other types of specialists if they find it necessary.

Chemists carry a wide variety of medications and remedies. You may find Chinese Deers Horn slices next to Arnica 6X (homeopathic), next to *Fluidextractum Valeriana* (medical herbalist) next to Brahmi Pills (Ayurvedic) next to Potter's Anteffect Pills (herbal OTC) next to Listerine. Take your choice, and get your script for Ampicillin filled while you're at it.

What this means, in the United States at least, is that in order to have such choices, you must do more for yourself, take on more personal responsibility for knowing some of the things that might help you, yet know when to seek the physician's aid. These plant medicines can have such a sensible place in self-medication or preventative treatment, especially when American medicine allows us so little access to the trained therapists of disciplines other than standard practice medicine. Thus, this second manual of Western plant medicine is offered for the intelligent use of the intelligent reader, both as a basic reference for gathering, preparing, and using herbs, and for understanding their implications. Readers who are physicians, pharmacists, chiropractors, naturopaths, nurses, and therapists may find it a useful starting point for further endeavors.

Herbal remedies represent far more than a wholistic fad or a total rejection of traditional medicine. They help to fill the overwhelming void between health and acute disease, a void left when the medical profession discovered penicillin and began its romance with the concept of "miracle drugs." I have known perfectly intelligent physicians whose sole regularly used reference manuals were the *Physicians' Desk Reference* and Goodman & Gillman, both drug manuals. Their patients have come to expect, and receive, prescriptions as their only therapy. At best doctors might have had a year of pharmacology in medical school (probably six months); the actual experts on drugs and their interactions, however, are the pharmacists who count out the pills. Yet how many doctors ever ask a pharmacist for advice? Very few, I imagine.

Drugs can save lives in serious, acute diseases and may alleviate or even cure some chronic problems, but they are very poorly suited for the vast majority of chronic and minor ailments. With their many and not-so-subtle side effects, they often hinder the innate ability of the organism to heal itself. Iatrogenic (drug caused) diseases affect most Americans at one time or another, yet a man or woman has innate disease defenses of incredible complexity, far surpassing the totality of all drug therapies. A disease that has run its course, with the body's being allowed to heal itself in its own fashion and chronology, is far less likely to return as a chronic ailment. With many patients seeking a pill and a pat on the back for self-limiting disorders and often unwilling to accept responsibility for their own healing, overworked physicians will prescribe drugs as a knee-jerk therapy, often circumventing and interfering with the body's responses.

This may allow the imbalance to occur at a later date, as the full cycle of defenses has been prevented, until the imbalance sets in as an ingrained habit called chronic disease. Although the life span has been greatly increased in the last hundred years, this is generally the result not of drugs but of preventative medicine and sanitation. We no longer defecate in our water supplies and we wash our hands before delivering babies. The more virulent micro-organisms that can penetrate the defenses of healthy individuals have been systematically controlled . . . at least in the industrialized world. What remains is a miasma of degeneracy diseases—the final states of the chronic imbalances, diseases that drug therapy and surgery have been singularly unable to deal with.

In contrast to the single-mindedness of drugs, the biochemical compositions of plants are varying, complex, and subtle. This generally makes them inappropriate for the chronic disease that becomes acute or for a seriously flawed defense system. At the same time, they are composed of fairly predictable groups of substances that our bodies have long since evolved methods of dealing with. In fact the very act of excreting some plant substances is often their mode of effect, since they are usually of little use as food or building materials. These predictable means of excretion explain why some herbs act to stimulate urine production, sweat and sebaceous secretions, lymph flow, and liver function.

The herbs in this book will often supply palliative relief (to ease the discomfort, as with a sedative, laxative, etc.) like many drugs, but they will do this without interfering with normal healing processes or having the residual toxicities of many drugs. Others will react in an opposite fashion, acting to stimulate and even temporarily aggravate the healing processes of the body. The concept of stimulating a fever seems to run counter to much accepted medical tradition, but when a fever is viewed as the only appropriate defense response at that stage of an infection, and when such an herb will also increase sweating, which brings heat outward as well as cooling the skin by evaporation and speeding excretion of waste products, this approach seems more rational than suppressing the fever with salicylates. In fact, many "diseases" and their symptoms are in reality perfectly appropriate body defense responses and should not be depressed at all. A simple inflammation is the reaction of the tissues to an infection or irritation, and results in quicker tissue repair or the isolation of the invader. Instead of lessening the inflammation through anti-inflammatories, it would be more sensible to stimulate it through rubifacients, moist heat, and a systemic vasodilator such as Indian Root, ginger, or capsicum.

This stimulation of defense responses, termed a *healing crisis* in naturopathic medicine, is the cornerstone of most alternative healing therapies, although herbs are only one way to bring it about. Plants that stimulate such responses are termed *alteratives* in herbal practice.

In such a use of an herb, however, there is the presumption that the person is in basic good health with reasonably attuned and appropriate defense mechanisms. They should be used with much caution for young children, the elderly, and individuals with chronic poor health or serious hereditary imbalances, as such people may overreact. In any case, common sense should be used when a defense reaction seems excessive—the body is not always correct in its assessment of the problem, or in the magnitude of its response. There are no fixed methods to apply to the human predicament, there is no single all-pervasive rule to follow, since medicine is not a science but an art. The therapies or uses recommended in this book cannot be treated as Holy Writ; they are simply sensible guidelines and majority effects.

The herbs I discuss will have no beneficial effects at all for some people—nothing works for everybody. Such factors as race, diet, sex, and even the time of day have many and complex effects that are poorly understood or even ignored, and the ingestion of biochemical substances to promote healing is only a small part of the whole.

Certainly herbs alone cannot bring relief to individuals whose very life style may be the cause of their illness. A person who has gradually developed a gastric ulcer and insists on coffee and doughnuts for breakfast, a ham sandwich and beer for lunch, a sullen stress-filled dinner, and endless rehashings of daily traumas and frustrations while sleeping cannot obtain more than the slightest relief from Algerita or Añil del Muerto. That is beyond the help of a mere herbal tea and beyond the scope of this book.

The avowed function of any healing therapy should be to enable the organism to heal itself by strengthening its defenses or, failing that, to caulk the cracks. The ways are legion, from the various biochemical approaches (allopathic, homeopathic, and herbal medicines) to naturopathy, chiropractice, rolfing, polarity therapy, meditation, rebirthing, organized religions, faith healing. The native and naturalized plants I have listed are a valid part of this healing circle.

If you like lists, here is one to keep in mind when using this book:
1. Be sure of the plant you are picking.
2. If the herb makes you sick, take less or throw it away.
3. If it doesn't work, use more or forget it entirely.
4. Trust your own judgment above all.
5. That which stimulates can irritate; that which helps can hurt.
6. If you don't get better or get worse quickly, call your doctor.

Medicinal Plants of the Mountain West (MPOMW), published in 1979, was received lukewarmly by professionals in the fields of ethnobotany, pharmacognosy, and natural products chemistry. Conversely, physicians and botanists liked it a lot, and those who used it the most were the intelligent, outdoorsy generalists I really wrote it for. This pleases me.

This book contains sixty or so new plants and a handful that were discussed (usually as other species) in *Medicinal Plants of the Mountain West*. The majority of the new plants were unfamiliar to me when I began assembling the experience and practicality for a desert and canyon book, and it has taken awhile to get better acquainted with many of these plants. In the process I have grown very partial to them, partial to the minimalist nature of desert life, and far more understanding of the riches and intricacies to be found in the Great Basin and Southwest deserts and their canyons.

To understand, gather, use, and research many of these plants and to know them as well as I knew those in the first book has taken some effort. Some of those I had expected to write about are, in reality, either not very useful, too duplicative in use, or too rare. Some rare plants, such as Elephant Tree and Chaparro Amargosa, are such good medicines, so little affected by judicious gathering, and so hard to find that I have included them here. Both are abundant in northern Mexico. Some common and uninspiring plants, such as Prickly Pear, supply a large array of therapeutic uses. Some dreadful grunt-weeds, like Puncture Vine and Milk Thistle, are clear and useful medicines, widely employed in the pharmaceutics of other countries. It has been possible to include herbs useful for a variety of purposes, found in all types of environments, from the city vacant lot; to roadsides; mesas; lovely, moist canyons; dry, cold, desert uplands; low, sweltering deep deserts; even scrabbly, overgrazed junk-places.

When I go out to buy a book on building passive solar homes, or on the best places to hike in the Chiricahua Mountains, or on making my own low-tech high-comfort furniture, I expect the writer to know so much about the subject that he/she knows what less knowledgeable me *needs* to know, able and willing to share opinions, understandings, pressure points, gestalts, processes, procedures, and proba,ilities. This is called the "Whole Earth Aging Hippy Post-Grad Dropout Approach to the Low-Tech/High-Tech Back-to-the-Earth" school of writing.

With this in mind, I have written for *you* the kind of book *I* would like to find, were I searching to know about the medicinal plants of the desert and canyons of the West.

FORMAT EXPLANATION

COMMON NAME: I have listed the plants under the most common name used in the West when the herb is addressed as a medicinal plant. An herb with a Spanish name such as Yerba Mansa may be listed in field guides as Lizard Tail or Swamp Root, but its only remedy use comes from Hispanics in Arizona, New Mexico, and Mexico; Echinacea is the plant often sold in nurseries as Purple Coneflower and has enjoyed much folk use as a medicine in bygone years as Black Sampson. But these days when it is referred to as a botanical drug it is called Echinacea.

LATIN NAME: Although the original intent of Latin naming was to create a universal, inviolate nomenclature, that is hardly the case anymore. I have tried to list the most commonly used species names from the plant manuals of the West or, where two names are equally employed, to list both. For those unfamiliar with Latin names, the first name is the genera (like Chavez); the second name is the species (like Maria); the last name is the Natural Order or the superfamily it is found in (like "Spanish-speaking New Mexican").

OTHER NAMES: These include some of the more common English or Spanish names that might be encountered, as well as the growing number of alternate Latin names.

MAPS: Unlike many mountain regions, where the ecologies of high country life-zones are often very similar and the same botany might be found in the mountains of five adjoining states but in stands hundreds of miles apart, our desert regions are distinct and highly idiosyncratic. The plant communities of the Great Basin, the Sonora/Arizona/Colorado Deserts, the Chihuahua Desert, and West Texas are rather dissimilar. I felt that, for the majority of plants, it would help to have approximate distribution maps. Some plants, such as Chickweed, Canadian Fleabane, and Puncture Vine, are commonly occurring weeds and need no map.

APPEARANCE: For identification purposes, the ideal circumstance is to go into the field with an experienced person. Most universities, junior colleges, and herbariums have classes or conducted plant walks, as well as extension and adult education programs. If you have never attempted to identify plants from manuals, even as unorthodox a manual as this one, it is always better to break the ice in such a manner. There are native plant societies in many areas as well as many individuals in the West that give herb walks.

Lacking such resources, or having some background other than strict botany, you will have little difficulty with the plant descriptions in this book, since the botanical jargon has been kept to a minimum. If you have training in botany you will probably tear your hair out, as there are instances where thirty words have been necessary to describe a plant detail when one technical word would have sufficed.

In general, where there are a number of species I have tried to include the pre-

1

dominant characteristics of the whole group, emphasizing distinctive peculiarities—those most likely to be noticed. The growing conditions and local varieties may greatly alter certain gross aspects, such as height, density of foliage, color of leaf or flower, scent, and the like. The illustrations are as typical of the overall appearance as possible, but many plants have the infuriating habit of adapting to their locality and not to this book. Remember that some plants maintain a uniform appearance wherever they are found, while others change subtly from meadow to meadow. In the Willow family (Salicaceae), Poplars maintain this uniformity, with Aspens never changing appearance from Alaska to Mexico, Pacific to Atlantic. The species, subspecies, and varieties of Willows, however, number well into the hundreds, with many local adaptations having characteristics of several types. A "species," therefore, as well as a genera, is often only an approximation, a series of fine lines arbitrarily dividing similar plants. With fewer plants and more botanists, certain groups of adaptive plants are fair game for master and doctoral "review." It is the accumulated weight of these "clarifications" that can cause the plethora of Latin names for the same plant.

HABITAT: This has to be approximate, since distribution of plants, their relationship to each other and our species, and distinct changes in plant ecologies from year to year make it impossible to define absolutely. A few of the introduced plants, particularly those with barbed seeds, can be encountered almost anywhere. Plants liking roadsides can move gradually or suddenly, their seeds carried long distances by road maintenance machinery; flood runoff and wildlife seed dispersal can rapidly extend a particular plant's range. The many desert plant herbariums in the West, usually at universities and botanical gardens, are an excellent beginning in locating plants of a more obscure nature. Noxious weeds, such as Puncture Vine and Milk Thistle, can be located by asking at your county agricultural extension service or at land grant and aggie colleges. Listings of more than twenty years past can be misleading; the wildlife of the Imperial Valley, Phoenix, Albuquerque, El Paso, and the Texas Panhandle areas has drastically changed (ceased, to be honest) in the last two or three decades. In urbanite areas, at least, go for more recent samples.

CONSTITUENTS: These are those substances, either complexes or single substances, that can be found in the plant discussed or in related species of a similar biotype. I have been rather meticulous in this regard, trying to mention those known substances in a given plant type that have some bearing on its therapeutic use (or might, or could, or who knows). I have made no effort to include some of the more passive substances most plants contain, such as starches and such. In some instances, the constituents are not yet known. If this information helps a few of you readers and users with technical backgrounds, fine; if the names impress some of you and make your left brain respect the plant more, fine; if they hold no meaning or interest for you, that's also fine; I feel all three ways on different days of the week.

CULTIVATION: Wild plants are tricky. If you are serious, get a soil-testing kit, test for pH and such when getting wild plants, and duplicate those conditions in your garden. Land-grant universities in all states will analyze soil for only a modest fee. Further, if a plant likes slopes, prefers open meadows, or cloisters under bushes, duplicate that in your garden. Learn to observe the growing conditions of the healthy plants in the wild. If you are going to transplant roots, get them after all the above-ground plant has died back, and get *all* the root. If the plant grows in the high mountains and you are gathering mature seeds, pack them in native soil, put them in your freezer, and alternate freezing with defrosting several times until the seeds start to germinate. In some plants this can take a year or more, while others will start to sprout in your freezer. Yet other plants will simply rot no matter what you do. As a rule of thumb, common plants are easiest to cultivate; uncommon ones are probably scarce because of their adaptation to very specific and narrow conditions. Members of the orchid family are always a lost cause, and those of the lily family may demand heroic efforts, so good luck.

COLLECTING: Make a point of picking only plants growing in prime locations. Individuals with many insect holes and obvious poor health are probably located at the extremes of their preferred growing conditions and may also have distinctly atypical biochemistries. Always check around after you have located a needed plant. There may be a whole field of it over the next rise or around the bend in the road. On the other hand, they may be the only two plants in the whole valley—and should be left alone. Furthermore, a plant common in one state may be an endangered or protected plant in another, so check first, if in doubt.

Certain conservation practices are always necessary. If a plant grows in large stands, never pick more than a third of the plants. If it is a larger, solitary bush or tree, never pick more than a third of the foliage or twigs and preferably from the borders of the plant, leaving the older central growth to regenerate outwards. If you are digging roots, dig no more than half of the immediately visible plants and the largest of the group, leaving the younger to grow and reseed. Fill up your holes if they are deep; an arroyo full of potholes is an invitation to erosion. The desert seems endless, the canyons protected and omnipotent, but both are delicate ecologies. With desert plants, the blooming is short (usually after rains), the seasons more a matter of local conditions than the calendar; so you need a good sense of their habits. Urban weeds are variable as well, with many manifesting second seasons, second growth in moist summers, fall monsoons. Rule of thumb: smaller ephemerals and urban weeds sprout any damn time they wish (or can); larger plants, like Mesquite and the Cacti, follow rather rigid timetables, often going through certain cycles, varied only by as little as a week, from year to year.

COLLECTING AND DRYING METHODS:

Method A. This is the preferred method for drying aboveground foliage that forms distinct stems. The stems are cut below the lowest green leaves or bark, bundled facing the same direction, then bound one or two inches from the cut ends with #31, #32, or #33 rubber bands (one-eighth-inch wide). Twine or wrapped wire may be used, particularly for drier-area botanicals that will shrink little. If the plants are unusually dirty or hairy-viscous (such as Camphor Weed or Yerba Santa), they may be washed under cool water, squeezed gently, rinsed, shaken drier and fanned out, and hung to dry. Washed bundles must be smaller in circumference to prevent spoilage. In reality, few herbs need to be washed if they are gathered clean and, where possible, away from roadsides. Many roadside herbs will absorb exhaust substances and should be cleaned.

Herbs must *never* be dried in sunlight; instead, they should be hung from hooks or nails in shaded areas with adequate circulation until both the top and bottom of the bundle are brittle dry. Many of our drier plants may be bundled, several bundles placed loosely in a large brown paper bag, and tucked into a dry, dark corner until desiccated; this is not a particularly applicable short-cut when the bundles are made of moister plants such as California Mugwort, Passion Flower, and the like.

After drying, they may be stored in a variety of ways, but the bundles are usually broken down or chopped first. For most herbs in blossom, the flowering half of the bundle is chopped into regular half-inch to one-inch segments with pruning shears or kitchen shears. The remaining half of the stems are stripped of leaves and the bare stems discarded. Many of our desert and canyon plants are atypical in growth, parts used, and the like, and, unlike the previous manual, I have generally dealt in greater detail with collecting methods for many of these plants.

Canning jars and cleaned reused jars with tops are the optimum storage containers. Be sure to label the contents and, if you have suppressed Virgo tendencies, you might even methodically label the date and picking location for future reference. Coffee cans or even plastic bags are also appropriate, but paper bags should not be used. Storage should be in a cool, dark area. If your area or house has grain weevils (little brown nasties that live in dried flour bins), then a pinch of diatomaceous earth should be added to the herbs and dispersed by shaking. This is a problem in moderate and high-humidity areas, particularly in stored roots and flowers. Most freshly picked, coarsely chopped herbs will last a year at full strength. After two years they should be discarded.

Method B. This is a simple method for small roots, seeds, fruit, and resins. Use the lower half of cardboard cartons that beer cans and soda cans are shipped to grocery and liquor stores in. They are nearly always sliced along the sides and the cans stacked in the half-box flats, discarded after the canned drinks are sold. They are clean, easy to get, and many plant materials can be loosely placed along the bottom of the flat, another stacked on it at a right angle, filled, and another flat placed on it at the same angle as the first, and so forth. When the botanicals are dry, they should then be stored as described in the previous paragraph. Many larger herbs, such as Red Root and Ocotillo, need to be cut into smaller sections while still fresh; this is an ideal way to process your herb on a weekend camping trip. Simply carry the flats back with you and continue the drying in the same flats at home.

STABILITY: This defines how long you can expect a gathered plant to stay reasonably strong or what characteristics it must maintain to still have potency.

If not discussed for a particular plant, presume the rule of thumb; strong up to a year, toss if two years old.

PREPARATION:

Teas: Cold Standard Infusion. Suspend 1 part (by weight) of the herb in cloth or paper towel in 32 parts of water at room temperature for at least six hours, preferably overnight, squeezing out the excess tea from the herb packet when finished. It is best to moisten the dry herb first before suspending it; gravity does the rest. The substances that dissolve in the water are heavier than water, sink to the bottom of the jar, and set up a slow convection current. The water containing more solubles is heavier than the water containing less solubles, so there is always a rise towards the top of weaker tea, always a draw down from the suspended herb. A very efficient method and, as it uses no heat, the least altering to plant constituents of any tea process.

Standard Infusion. Boil 32 parts of water, remove from heat, and steep 1 part by weight of the herb in the water for one-half to one hour, depending on density of the botanical. Strain and add enough additional water to the tea (pouring it through the herb in the strainer) until the original 32 parts is reached.

Strong Infusion. Boil 20 parts of water, add 1 part herb, proceed the same as above, adding enough water (if needed), through the marc, to make 16 parts of tea. Twice the strength of the standard infusion.

Strong Decoction. Bring to a boil 1 part (by weight) of the herb in 20 parts of cold water. Continue boiling for 10 minutes, remove from heat, and cool until body temperature. Strain, and add enough water (if needed) to make 16 parts of tea.

Weak Decoction. Same as above, but use ½ part (by weight) of herb. *Do not make more than a day's worth of tea at one time.*

Eyewash. Wherever the making of eyewashes is mentioned, remember two things especially: make the tea with isotonic water (neutrally saline) by using clean or distilled water (1 quart) combined with a slightly rounded measuring teaspoon of table salt (½ teaspoon per pint, ¼ teaspoon per cup); secondly, make a fresh batch of tea each time you use the eyewash or, at the very least, discard the unused eye solution after five hours. You want clean stuff in your eyes even when they are all right; sore eyes can be especially susceptible to infections.

Salves: Method A. This is an efficient method of making salves out of plants that are not very oil soluble, by using alcohol as an intermediate solvent. Grind up the herb first, using a blender, grain mill, or mortar. (In order of ease . . . mortars are difficult to use, at least for me.) Take 1 part by weight (2 ounces, as an example) and moisten with 1 part by volume (2 fluid ounces, to continue the example) of pure grain alcohol (95% ethanol) or, if not available in your state, 90% isopropyl rubbing alcohol (most rubbing alcohol is 70%, so you might have to get your pharmacist to order it). Moisten the herb with the alcohol by

5

hand or spatula until evenly dispersed, and place in a covered container for one-half hour to macerate. Place the alcohol-moistened herb in a blender and pour 7 parts by volume (14 fluid ounces in this example) of vegetable oil over it. Blend everything for several minutes until the glop is warm from the agitation. Place a clean cloth inside a large strainer or colander and set colander in a bowl. Pour the oily-boozy stuff into the cloth, let it strain for awhile, fold the edges of the cloth together, and squeeze as much as you can out of it, discarding the marc. If you don't mind a little alcohol scent, proceed forthwith to the salve stage; if the presence of about 12% alcohol is bothersome, let it sit in a broad, flat bowl on the windowsill for eight or ten hours, or place the bowl in front of a fan for several hours (in a well-ventilated room), then take the steeped oil and make the salve. To harden the oil, dissolve slowly 1½ parts beeswax (3 ounces in this case) over a low heat in the steeped oil, *just* until it is dissolved. Pour it into containers, wait until it hardens, and screw the lids on. Try not to heat the oil and beeswax too much, just enough to melt it together; if you want to play it safe, heat them in the top of a double boiler.

Method B. This is for those herbs that are readily soluble in oil. Grind 1 part of the herb, combine in a blender with 7 parts of vegetable oil until warm, pour into a jar, and set aside for a week. Pour it back in the blender, blend until warm, strain, and proceed with 1½ parts beeswax as in Method A. *Note:* If you don't wish to bother with the salve texture, or prefer the idea of just oil, delete the beeswax and heating step and pour the steeped oil in a jar. Salves oxidize (or form lipid peroxides) and turn rancid very slowly, and any old vegetable oil works fine. If you want to keep it as an oil (Method A or Method B), you need to use olive oil or sesame oil, as these form peroxides very slowly and take much longer to turn rancid.

Poultice: A hot, moist mass, consisting of a base (slippery elm, comfrey root, clay, flax seed, kudzu, bran, etc.) and one or more active substances (mustard seed, ginger, anemone, etc.) and placed on any part of the body, held (usually) between two pieces of muslin, and changed when cool. This aids pain, congestive inflammation, and tissue damage, as well as speeding absorption into the poultice of waste products, protein metabolites, and free amino acids.

Tinctures:

Method A. Simple and elegant. Take 1 part (by weight) of the fresh plant, just gathered and rinsed, chop it up into small pieces, place in a clean glass jar with a good lid, cover the chopped herb with 2 parts of 95% ethanol (grain alcohol like Everclear), screw the lid on after making sure the herbs are compressed enough in the jar that the alcohol comes up to the top of the chopped herb, and set it aside, untouched, for 7 to 10 days. The alcohol dehydrates the cells of the fresh plant, drawing out all the plant's substances into the fluid until, by the time you drain off the finished tincture, it has a deep, radiant color and the herb is yellowish, dusty white, and exhausted of its constituents, color, and essence. Pure grain alcohol is available from liquor stores in all the western states except

California. This tincturing method needs 190 proof to work properly; 80 or 100 proof (40% or 50% alcohol) is much inferior in the extraction capabilities, so Californians are urged to go to Arizona, Nevada, or Oregon to make fresh plant tinctures. *Puro de Caña* (pure grain alcohol) is always available in Mexico, is far less expensive, and you may bring across a single, one-liter bottle per month per person, duty free. I suppose Californian tincture-makers could pay duty for additional bottles without breaking state laws, but you'd better check with the U.S. Customs office to find out specifics. Mexican *farmacias* are a good source of Victoria brand *puro de caña*.

Method B. The simpler approach is the classic one of grinding the herb, weighing the coarse powder, adding the solvent, putting the mixed gloop in a canning jar, and shaking it twice a day for two weeks. This is a fine and time-honored method, producing a full-strength tincture, and is considered a proper alternative method in pharmacy. To outline by example, here it goes. Grind and sift four ounces of *Echinacea angustifolia* dried root. You need to make a 1:5 tincture with 70% alcohol and 30% water; when you make a *steeped* tincture (Method B) you add the menstruum (70%/30% in this case) that has five times the volume (20 fluid ounces) than the herb weighs (4 ounces) and, after two weeks, what you can squeeze out is what you have. How much alcohol? (20 oz. times 70% is 14 oz. alcohol). How much tincture will you end up with? (probably 11 or 12 oz., with good wrists). If you want the most out of the steeped herb, you need a mechanical squeezer, such as a Norwalk Juicer or a fruit juice press. With a mechanical squeezer you will be able to express as much as 15 or 16 ounces of tincture out of the 20 ounces of liquid you put with the 4 ounces of ground herb. Remember, you put *in* the five parts of menstruum (solvent) . . . what you get *out* depends solely on the efficiency of your squeezing methods.

The other, preferred method is called percolation and is a neat, low-tech method that is the official procedure for tincturing but needs a special percolator glass and some hands-on practice that is poorly described but easily *done*. The whole procedure is explained, rationalized, and illustrated impeccably in any edition of Remington's *Pharmaceutical Sciences*.

MEDICINAL USE: Several considerations must be kept in mind when using herbs to remedy a physical imbalance or disease. Acute illnesses, those with quick onset, strong symptoms, and a self-limiting nature, should be treated simply, using one or two herbs. The purpose of botanicals here is to give comfort, speed defense reactions, and limit and define the course to prevent complications or prolonging. Common sense is paramount, since the remedies may not be sufficient or may cause an overreaction. The use of salicylate herbs (Birch, Poplar, Willow, etc.) may turn a fever into chills. Conversely, stimulating the fever, as with Elder or Yarrow, can prove excessive on occasion. Any reaction which itself denigrates or impedes the body's strength must be avoided, since one of the main validities of proper herb therapy is to aid and augment defense responses without hindering them with toxicities. Since the pharmacology of most herbs is so diffused, they are rarely focused enough to supplant or sidestep body defenses in the manner of some drugs. Although these same drugs will usually have distinct secondary toxicities, they often serve valid semiheroic functions where an individual has failed to regain internal equilibrium. Excessive quantities of an herb sufficient to cause a toxic reaction simply compromise basic health without the synthetic defenses offered by some drugs.

For chronic illness, more complex combinations are usually more effective than single remedies, with small regular doses preferable to erratic large portions. The dose should be small enough that no overt symptoms are produced. To facilitate the effectiveness of long-term maintenance remedies in chronic disease, a simple formula may be used. As an example, a person has a chronic pulmonary weakness, with a history of asthma as a child and a tendency to have most common viral infections settle in the bronchials. This formula would go as follows:

1. SPECIFIC: (Yerba Santa, Grindelia, Inmortal, or Pleurisy Root)
2. LAXATIVE: (psyllium seed, Dandelion Root, etc.)
3. LIVER STIMULANT: (Burdock, Yellow Dock, Toadflax, etc.)
4. DIURETIC: (corn silk, Cleavers, etc.)
5. LYMPHATIC: (Red Root, echinacea, etc.)
6. NERVINE: (Skullcap, passion flower, etc.)

In conclusion, #1 should be an herb or herbs dealing with the chronic problem and should be the largest component. The rest, #s 2–6, are "satellite" herbs to facilitate and diffuse. None of the nonspecific herbs should be present in palpable quantities, that is, the laxative should not have a pronounced laxative effect, the nervine should not be strong enough to cause any drowsiness, and so forth. The rest of the formula is only to aid in making changes that can occur as a result of the specific. Unlike acute disease, the course of chronic illness is slow and usually submerged; therefore, treatment will always take weeks or months.

Further, most chronic problems not of direct genetic cause will generally derive from imbalances in life style, diet, emotional or spiritual instability, and stress. The end result (chronic disease) can be considered as a negative habit of body function or response. The best time to instigate therapy on any or all of these levels, including herbal, is during periods of change. For some the spring or fall season is most auspicious; for others it can be moving to another place, leaving a relationship, having a child, or changing jobs, religions, or diet.

The combining of drug and herb therapy can be useful, useless, or disaster prone and is far too unpredictable to deal with in great detail. An herb such as Alfalfa, with virtually no systemic effect other than as a source of soluble nutrients, is a useful adjunct to drug therapies, but otherwise it is safest to leave each to its separate realm. Some specific horrors should be mentioned, however. Like aspirin, the salicylate herbs should never be combined with anticoagulant drugs. These include Birch, Poplar, Willow, and probably even members of the Ericaceae order such as Blueberry, Pipsissewa, Pyrola, Manzanita, and Uva Ursi. Herbs with pronounced sedative effects, like their drug counterparts, should not be consumed with alcohol . . . or their drug counterparts. More complex and unpredictable drug approaches such as anticholinergics, adrenergic blocking agents, and the like should be taken under the closest supervision and not combined with any herb, since one can create a witches' brew of side effects. Herbs containing coumarin may be safe as teas (Sweet Clover, woodruff, etc.) but can become frankensteinian combinations with some drugs. A plant high in tannins will prevent or slow down absorption of drug substances or even precipitate them out completely. At the same time, laxative herbs or those affecting liver function may seriously alter the predictability of a drug action. Much study has gone into the fate and absorption rates of drug therapy, and doses are set based on normal metabolizations, with herbal laxatives and liver stimulants only interfering.

The time of day and mode of use affects the strength of reaction to many herbs. Sedatives and laxatives work best when their use coincides with normal patterns of sleep and defecation. A bitter tonic or stomachic works best when taken shortly before meals or predictable discomfort. Potential irritants, such as Bayberry or cayenne, should be taken on a full stomach, herbs meant to work quickly, on an empty stomach. An herb taken for recurring symptoms that give advance warning (such as migraine headaches) or are part of a predictable stress (such as insomnia or hangovers) works better when taken *before* the discomfort has ripened.

In dealing with young children a great deal of caution must be used, since body defenses can quickly prove inadequate, and the speed at which an ailment can become dangerous is often foreshortened. In young children and infants, the speed of infection may not be quantitatively different than in an adult, but an organ or tissues can be compromised much sooner because of the considerably smaller volume of resistant tissue. One should be especially cautious in totally relying on home remedies for small children when the sickness is febrile (feverish), eruptive, or involves diarrhea or the eyes, ears, or mouth. Also, any lung infections should be approached conservatively. The most fanatically devout follower of natural healing methods should still take an infant or young child to a physician when there is any doubt at all, since the course of such disease can be quick, volatile, and unpredictable. On the other hand, children respond very well to the simplest, most benign herb remedies. Seldom is anything stronger than Vervain necessary to bring them both palliative and substantive relief. Doses, of course, should be a half or a third as much.

Similarly, certain modifications should be used in treating the aged. They are usually more sensitive to herbs and drugs, and smaller quantities should be used, from one-half to two-thirds as much. Special care must be taken when an herb has a nauseating or cathartic effect. A safe quantity of Lobelia under other circumstances can induce a depressing, clammy nausea in an aged person; an energetic laxative may produce painful cramps and irritate the colon or small intestine. The equilibrium of health is often more delicate in the aged, with only small changes causing great discomfort. Most illness of age is the direct result of chronic disease, and herbal medications should be given using the formula under chronic disease therapy above.

Rare diseases *are* rare, and most discomforts and illnesses can be helped with herbs. Still, the ideal circumstance is to know (or be) a physician that will allow for the validity of herbal medicine yet act as a screen for more serious problems. Lacking that, get to know your body and use common sense. Even though a drug therapy may not always be the best approach, the best diagnostician for the nuts-and-bolts mechanistic problems of the body is a physician, the best judge for the drug therapies is a pharmacist.

ACACIA

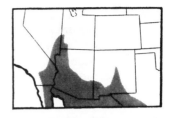

Acacia greggii, A. constricta, and
others   Leguminosae

OTHER NAMES: Catclaw Acacia (*A.
greggii*), Whitethorn (*A. constricta*),
Huisache, Cassie, Timbe, Vinorama

APPEARANCE: The two species listed above are our dominant Acacias, so I will limit myself to describing them. Catclaw (*A. greggii*) is a small tree or shrub, eight to fifteen feet high, though capable of forty feet or more in ideal locations. The "catclaws" are like rose thorns, broad at the base and curved backward, spaced irregularly along the branches. It can be confused with Catclaw Mimosa, a plant that looks similar but has white or lavender, tufted-button flowers; Catclaw Acacia has yellow, cylindrical spikes. The flowers and leaves of this plant resemble Mesquite, but the thorns of the latter are straight, and the constricted three- to five-inch pods of the Acacia split after maturing, whereas Mesquite pods do not.

Whitethorn (*A. constricta*) has long, paired white spines at the nodes of each offshoot stem. In new spring growth, they may be as long as one and one-half inches . . . ouch. The flowers are yellow and form round balls about one-half inch in diameter. They mature into three- to five-inch-long skinny pods that are tightly constricted around a single row of seeds, splitting soon after. If the Acacia looks like this, but instead has two rows of seeds in an unconstricted pod, you have stumbled onto Huisache (*A. farnesiana*). Except in southern Texas, it is uncommon north of the border. There is also Fern Acacia (*A. angustissima*), which has long, graceful secondary leaflets, no thorns, and light pink one-half-inch round flowers that nearly form racemes. Their brown pods are two to three inches long and slightly constricted. Other occasional Acacias resemble the previous three species so closely as to need no further description. Only the Fern Acacia is thornless.

HABITAT: Found in all of our deserts except the central and northern parts of the Great Basin; up to 6,500 feet but most common from 2,500 to 4,500 feet. Fern Acacia is mostly in the southern part of Arizona; the others are widespread.

CONSTITUENTS: Leaves, pods, roots: anisaldehyde, benzoic acid, benzyl alcohol, butyric acid, coumarin, cresol, 7,3',4'-trihydroxyflavan-3,4-diol, leucoanthocyanidin, N-methyl-B-phenethylamine, N-methylpentathylamine, N-methyltyramine and tyramine; gum: same as Mesquite Gum.

COLLECTING: Gather the pods when still green, dry, Method B; the leaves and branches, dry using Method A, but over a paper, as the leaves often fall off while hanging; the longer distal roots, chopped into small segments while moist, Method B. The gum is gathered the same as Mesquite Gum. The flowers are dried, Method B.

STABILITY: Indefinite.

PREPARATION: The green leaves, stems, and pods are powdered for tea (standard infusion) or for topical application; the roots are best as a cold standard infusion, warmed for drinking and gargling.

MEDICINAL USES: The pods are used for conjunctivitis in the same manner as Mesquite Pods are used, and the gum, although harder to harvest, is used in the same way as Mesquite Gum. The powdered pods and leaves make an excellent infused tea (2–4 ounces of the standard infusion every three hours) for diarrhea and dysentery, as well as a strongly astringent hemostatic and antimicrobial wash. The straight powder will stop superficial bleeding, and can also be dusted into moist, chafed body folds and dusted on infants for diaper rash. It was widely used by Native Americans for treating the sore backs and flanks of their horses.

The flowers and leaves as a simple tea are a good anti-inflammatory for the stomach and esophagus in nausea, vomiting, and hangovers. It is distinctly sedative. The root is thick and mucilaginous as a tea and is good for sore throats and mouth inflammations as well as dry, raspy coughing. A nifty plant.

AGAVE

Agave spp.   Liliaceae
OTHER NAMES: Century Plant,
Lecheguilla, Maguey

APPEARANCE: Agave looks very similar to Yucca and is often mistaken for it, but it is a more robust plant with thicker spiney-edged leaves and a flowering stalk that, unlike Yucca, forms distinct armlike branches.

HABITAT: From 2,000 to 7,000 feet on mesa sides, limestone slopes, and bajadas. It is especially common in lower, cooler canyons; the lower and hotter the area, the more protection it needs. It is found in the Mojave and Colorado deserts, the southern Great Basin and the deserts of Arizona, southern New Mexico, and West Texas. *Agave virginica* grows eastwards into the Ozarks and Appalachian mountains.

CONSTITUENTS: Sapogenin, hecogenin, tigogenin, gitogenin, etc.

COLLECTING AND PREPARATION: The fresh leaf, tincture using Method A; for tea, throw a few fresh leaves in a paper bag and forget about them until they are dry, then chop them up. The root should be cut into small slices while fresh, and dried, Method B. The root tincture is Method B, 1:5, 50% alcohol.

STABILITY: The dried leaves and roots are very stable, however processed.

MEDICINAL USES: The fresh leaf tincture (¼ teaspoon in water) and the dried tea are good GI tonics, useful for indigestion, gassy fermentation, and chronic constipation. The root tincture is an effective antispasmodic, helpful for gas pain and colic, as well as being a good bitter tonic. Water retention from little physical activity or major shifts in the weather are relieved by the strong diuretic effect of the leaf tincture, and arthritis aggravated by changes in humidity and barometric pressure can be helped by the root tincture, ¼ teaspoon in water three times a day. Mexican dark tequila may work similarly.

OTHER USES: The fresh root is a decent shampoo; grate it into a bowl and add some hot water until it lathers, then strain it through a cloth and wash with the expressed juice.

SIDE EFFECTS: Constant use of the root for arthritis may interfere with some intestinal absorption; so don't use for more than a week at a time. The fresh tincture may be irritating to the skin and mouth of a few individuals; so, if in doubt, put some on the inner wrist. If it inflames, don't use it; use the dry leaf instead, for tea.

ALGERITA

Mahonia trifoliata, M. haematocarpa, M. fremontii, M. wilcoxii

OTHER NAMES: Desert Barberry, Oregon Grape, Agarito, *Berberis* genus, Yerba de Sangre

APPEARANCE: These are shrubs and small tree-bushes of the middle and upper deserts. The leaves are spiny-toothed and evergreen, the color (except in *M. wilcoxii*) is glaucous blue-green and the shrubs, although scraggly, are handsome and decorative. The many yellow flowers bloom in the middle spring, maturing into small bunches of round berries. *Mahonia trifoliata* and *M. haematocarpa* have bright red, juicy berries, *M. wilcoxii* has berries that resemble blue-black grapes, *M. fremontii* has yellowish-tan dry and hollow berries. Although I am applying Algerita to these similar plants, Texas purists may argue correctly that *M. trifoliata* is the species most frequently called by that name. Nonetheless, the main difference is that *M. trifoliata* has three leaflets, the others have five or seven. The only one of these that doesn't have the small prickly-leafed, large blue-green desert shrub appearance is the *M. wilcoxii*, which is a dark-green, large-leafed and open-foliaged moist canyon Algerita, more like a giant Oregon Grape than any of the others. All have yellow bark and roots.

HABITAT: *Mahonia trifoliata* grows from Tucson to the Pedernales River in the Texas Hill Country, south into northern Chihuahua, usually from 2,000 to 4,000 feet. *M. haematocarpa* has a similar range but grows in the southern Great Basin in northern Arizona, and grows in slightly moister climates, usually mixed with oak, Juniper, and Piñon, usually from 3,000 to 5,000 feet. *M. fremontii* is primarily a Great Basin plant, found in Nevada, Utah, northern Arizona, New Mexico, and western Colorado, usually in dry Juniper/Piñon deserts from 4,000 to 6,000 feet or higher. *Mahonia wilcoxii* is found in very different circumstances; the moist canyons between the mountains and the deserts of southern Arizona, New Mexico, and northern Sonora, from 4,500 to 7,500 feet.

CONSTITUENTS: The related alkaloids berberine, berbamine, canadine, mahonine, oxyacanthine, magnoflorine, oxyberberine, obamegine, aromoline, etc., depending on species, time of year, and even more esoteric factors.

COLLECTING: Gather the roots and stem bark from midsummer to winter. *M. wilcoxii* with its different growing habits has usable lower stems that differ little from the creeping roots and may be used as high up the woody stem as the pith remains yellow. The other three shrub species have intensely yellow roots but much less strength in the trunk and stem woods; so use only their bark, mixing it freely with the whole root.

STABILITY: Several years, at least.

PREPARATIONS: These Algeritas have tough, dense roots; split them with hatchet, chisel, or limb shears and cut them while fresh. The fresh roots, Method A or, if you have a grinder adequate to the task, dry roots, Method B, 1:5, 50% alcohol. The herb is water soluble; Cold Infusion, 2-4 ounce doses. The tincture, fresh or dry, should be used in 5-10 drop doses as a bitter, 15-30 drop doses as a liver or alterative medicine.

MEDICINAL USES: Algerita has three main functions: as a bitter tonic for impaired salivary and gastric secretions, as a stimulant to liver protein metabolism, and as an antimicrobial for the skin and intestinal tract.

As a bitter tonic, it has the simple, predictable effect of stimulating (because of its nasty taste) salivary and parotid secretions and, by reflex, hydrochloric acid and pepsinogen secretions from the stomach lining. Take it fifteen or twenty minutes before meals (5-10 drops of tincture). An easy rule of thumb is the following: if you have indigestion regularly, have teeth or gum problems, and a white or yellowish coated tongue in the morning, use the herb. If, on the other hand, your indigestion is accompanied by excessive salivation and the tip of your tongue is red, your stomach lining is probably irritated and inflamed . . . use Arizona Walnut or something like Hollyhock to soothe it. You don't need a stimulus to secretion but something to cool and soothe. The bitter tonic use (as well as the liver-stimulating doses) are useful for stomach problems of the chronic drinker (subacute, congested membranes) but not for the stomach problems of the occasional binge drinker (inflamed, hot membranes) . . . unles you wish to use it for a hangover treatment.

Algerita is a stimulus to liver metabolism of dietary and blood proteins and is helpful for those people who show signs of moderately or constitutionally impaired hepatic function. This shows as dry skin, poorly healing skin, and mucosa injuries, constant bad breath and coated tongue (especially in morning), difficulty in digesting fats and proteins or even an aversion to them, rapid shifts in blood sugar levels, a life-long preference for sweets, salads, fruit, and the like, and a history of food or environmental allergies. Try the tincture (15-30 drops) or the tea (2-4 ounces) in the morning and before retiring for at least two weeks.

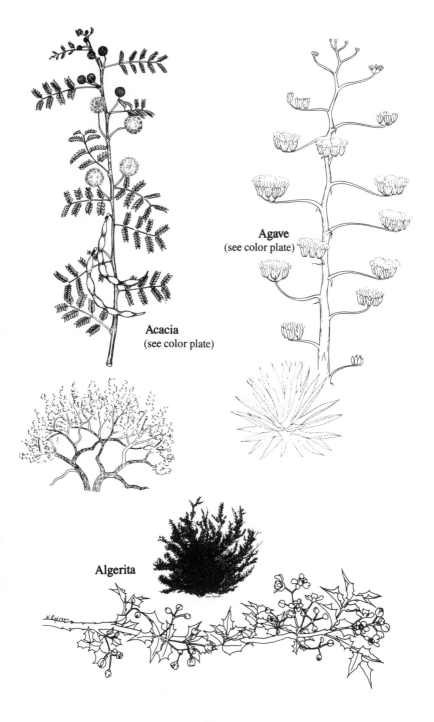

Agave
(see color plate)

Acacia
(see color plate)

Algerita

15

ANEMONE

Anemone tuberosa   Ranunculaceae
OTHER NAMES: Desert Anemone, Desert Windflower, *A. sphenophylla,*
A. edwardsiana

APPEARANCE: This is one of our most delicate and reclusive wildflowers, widespread but very difficult to find. The flowers, like Buttercups, have no real petals, only bright sepals, colored white to lavender red and eight to ten in number. The floral parts are many-stamened and yellow, maturing into a conical, silky seed head. The whole plant is from five to twelve inches high, with one or several long-stemmed leaves, divided palmately by threes. These resemble Parsley and often have a purplish stem and some purple on the leaves. The leaves on the flower stalk are similar, but clasp the stem and have fewer divisions. The root is a small, dark-skinned pinkish tuber.

HABITAT: Foothills, sheltered arroyos, and canyons in the transition between dry and moist terrain, from 2,500 to 5,500 feet. It is often found in association with Beargrass (Nolina) and Small-Leafed Sumac. It grows in the Panamint and Providence Mountains of California, through the northern Mojave Desert and most of the Great Basin, down through Arizona and into New Mexico as far east as the Davis Mountains of West Texas. Look for it to bloom from March (southern Arizona and California) to May (Utah and Nevada); if you miss the blooms, forget it until next year, as it disappears from the face of the earth.

CONSTITUENTS: Pulsatilla Camphor (anemonin, protoanemonin).

COLLECTING AND PREPARATION: Only the fresh plant is active, so plan to make a Method A tincture. The whole plant is useful, but gathering just the leaves and flowers allows the stubborn little tuber to grow next year. Wash and crisp up the plant before tincturing. *WARNING:* the fumes released when the fresh plant is chopped are highly irritating to the eyes and respiratory mucosa; chop in the open or with a fan to your side, blowing across the table.

MEDICINAL USES: This Anemone is interchangeable with the former drug plant Pulsatilla (*Anemone hirsutissima*) both in its constituents and its pharmacology. As the latter is virtually unavailable anymore in its proper form (the fresh plant tincture), this is a handy plant to have in our area. I have gathered both Pulsatilla and this species, and find ours equal to or superior to the official herb.

The specific indications for Anemone are insomnia, nervousness, and a generally agitated frame of mind, with a sense of gloom, distress, and feebleness. The person is not flushed or red-eyed, but wan and chilly, inflicted with what was once called noxious night vapours. In other words, if you are tired, overworked, but nervous and overwrought, have a rapid but thready pulse, want to injure someone (Anyone!), but are just too blown out to manage it, try some Anemone instead. In the Bad Old Days I used Anemone and lots of Passion Flower to help get friends down from hallucinogenic flights of abject terror, and still find it useful to calm down those who get anxiety reactions from marijuana.

Dosage is important; only use between 5 and 15 drops of the tincture. These small amounts slow and strengthen the pulse and respiration, improve the central nervous system blood supply and circulation, dilate the peripheral blood vessels, and allow a smooth, efficient fanning out of blood from the center of the body. Large amounts simply lower the blood pressure too much and over-suppress the sympathetic autonomic functions.

If any of you women get these kinds of irritable gloomy PMS symptoms, the older herbalists and physicians swore by Anemone for its value in "Menstrual Derangements," so use it. As a stodgy old male feminist, my inclinations are to belittle such potentially sexist concepts . . . but what do I know.

The Native American usage in Nevada and eastern California was to crush up the fresh plant and apply it for several minutes to sores and boils, removing it just before it blistered. This makes a lot of sense, as the aromatics of the fresh Anemone are strong counter-irritants and will help to restore blood to the injury and speed resolution. Further, anemonin is an acknowledged antimicrobial agent against skin bacteria.

CONTRAINDICATIONS: Pregnancy, bradycardia, circulatory disorders in general. Although often superficially appropriate, Anemone is a little too ragged for the aged.

AÑIL DEL MUERTO

Verbesina encelioides   Compositae

OTHER NAMES: Capitaneja, Golden Crownbeard, Goldweed

APPEARANCE AND HABITAT: This is a common roadside annual, up to six feet high but usually around three feet. It may fill whole vacant lots or cover miles of roads, and is so familiar in appearance that you may easily overlook it. The blossoms, up to a dozen on a plant, are bright yellow sunflowers, with yellow petals and yellow centers with a few brown specks. Our true Sunflowers are generally larger and have brown centers. The foliage is silvery green or grey and has a strong, fetid scent that is a little like rotting meat (thus the Spanish name for it, "Sunflower of the Dead"). The leaves are mostly clasping, with various teeth, flanges, and cuts that outline an overall triangular shape. It likes road-sides, waste places, and the like, growing up to 8,500 feet, but usually 3,000 to 6,500. It can be found in almost all of our area. Añil del Muerto is especially fond of unpleasant roadside ditches filled with beer cans and empty plastic oil containers. It can also be found, fortunately for us, along the edges of clean arroyos and hillsides.

COLLECTING: The whole (clean) plant in bloom, Method A.

STABILITY: The whole plant at least a year; the powder up to six months.

PREPARATION: The powdered plant mixed with water for a paste, or made into a salve, Method A.

MEDICINAL USES: Añil del Muerto is primarily an anti-inflammatory for redness and swelling of the orifices. The paste is applied directly to hemorrhoids, labial inflammations, and sore gums. If this is too inelegant for regular use (understandable), the salve is just as effective. For an especially good hemorrhoid ointment, see Formula #1 (page 137).

Generally speaking, this herb works best for those inflammations arising out of other acute conditions or major stress. For example, if you get a mild case of food poisoning, followed a day or two later by a hemorrhoid flareup. Or, if you stay up all night with pots of coffee and cigarettes to finish an overdue report and two days later get a fever blister or sore gums.

The tea, a cup in the afternoon or an hour before dinner, can be a major help in aborting an early peptic ulcer that may be instigated by job stress.

Finally, a hot cup of the tea will break fevers, inducing copious sweating, relaxation, and a mild laxative effect. This is the kind of plant that, when disrespectfully called "that damn weed!," makes me smile Mona Lisa-like to myself. We know better.

BUCKWHEAT BUSH

Eriogonum spp.   Polygonaceae

OTHER NAMES: Wild Buckwheat, Antelope Sage, Colita de Raton, Sulphur Flower, Skeleton Weed

APPEARANCE: These Buckwheats are an immense genus, particularly in California, Nevada, and Arizona; true cultivated Buckwheat (*Fagopyrum esculentum*) doesn't resemble them a whole lot; so don't take the common name too seriously. *Eriogonum fasciculatum, E. parviflorum, E. racemosum, E. deflexum,* and *E. wrightii* are perhaps the most common of the group, but please understand, this is a *big* genus. Munz lists 104 species in California alone, so let's talk about the genus in general, as you will likely run across it out there. Most Eriogonums have rosettes of basal leaves, long skinny flower stalks, and terminal puffs of flowers, often forming umbels. The flower colors are mostly white, pink, and sulphur yellow, with the very common California Buckwheat (*E. fasciculatum*) having new pink flower heads intermixed with older reddish-brown seed heads. Many of the various species are woolly or silvery, and most become reddish brown in the flowers, basal leaves, or stems as they age.

HABITAT: Buckwheat Bush grows around the edges of the terrain, seldom on the peaks, seldom in the rich valleys or canyons; usually in gravel, rocks, roadsides, along open hills and arroyos. They are found throughout our desert and canyon area, more sparingly in our northern and eastern sections.

COLLECTING, PREPARATIONS, AND MEDICINAL USES: Almost all of our information about using Wild Buckwheat comes from Native American and Mexican uses. The plants preserve well when dry, Method A or B; they are water-soluble, and help the body by slowing secretions, shrinking and soothing inflamed membranes. The tea is a good, reliable eyewash, and many of the California Indians use the tea for washing newborns. The flowers are especially diuretic. Although

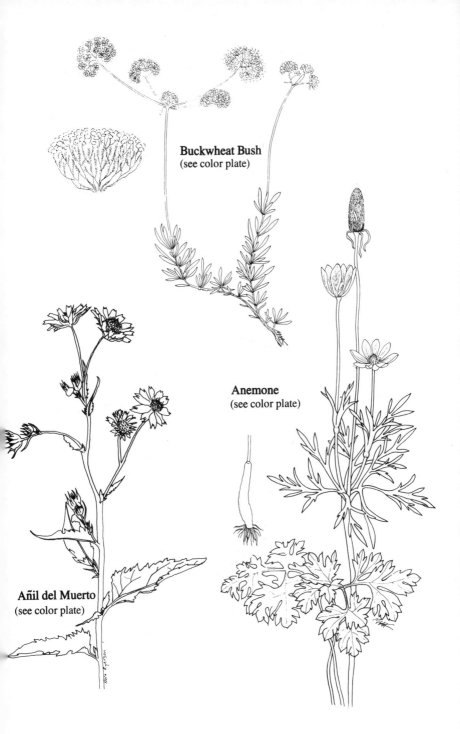

Buckwheat Bush
(see color plate)

Anemone
(see color plate)

Añil del Muerto
(see color plate)

19

astringent to the bladder and urethra, they are not irritating to the kidneys and have many uses in cases of cystitis and urethritis. The tea is also useful for premenstrual water retention and the fluid retention that often occurs in the last month or two of pregnancy. It decreases fresh spotting at the tail end of menstruation, and helps limit postpartum bleeding. The Cahuillas use the tea for dull, nagging pain in pregnancy, especially in the back and hips.

This herb's mild astringency makes it useful as a gargle for sore throats under any conditions and in any problem. Sick, slightly nauseous headaches that localize in the forehead and around the eyes respond well to the hot tea, especially if injudicious eating or drinking was the cause. These are homely and gentle plants, with mild, well-defined uses and no toxicity.

CALIFORNIA MUGWORT

Artemisia vulgaris, var.
douglasiana   Compositae
OTHER NAMES: *A. douglasiana,*
A. vulgaris, var. *heterophylla, A.*
californica, etc.

APPEARANCE: California Mugwort forms large, handsome communities of several to hundreds of plants. They are three to seven feet tall when mature by summer, connected by running rootstalks. The leaves are lance shaped, dark green on top, strikingly silver underneath, the lower leaves variously cleft, the upper leaves usually entire. Flowering heads are erect and arranged in dense or open panicles of the usual nondescript wormwood flowers. The floral stems themselves may be well over a foot long, and the many floral branches are leafy at their axils. The stiff stems are variously ridged for added structural strength and become green purple by summer. The whole plant has a strong but pleasant mugwort or sage smell and is bitter.

HABITAT: Coastal, from Baja California to British Columbia, east to the waterways of the Mojave and Colorado deserts; it does not make it into Arizona but is abundant from the Mojave River and the Imperial Valley north into western Nevada and the western edges of the Great Basin. It grows from sea level to 6,000 feet, always in relative moisture. The less the precipitation, the more shade it needs, until, with too little shade and too little rain, it disappears.

CONSTITUENTS: A number of sesquiterpenes and at least 22 related guainolides, caffeoylquinic acids, and chlorogenic acid.

COLLECTING: The whole flowering plant, dried Method A, stripped when dry and the largest stems discarded. The volatile oils are much of the active strength, so store the herb as whole as practical and in airtight containers; crush the leaves and flowers before using. For its bitter tonic and liver functions, a cold infusion is preferred; for other uses, prepare hot.

STABILITY: As long as the dried herb maintains its strong scent when crushed it is useful. This usually means from one to two years, depending on how well it is stored.

PREPARATION AND MEDICINAL USES: As with Silver Sage, the cold standard infusion (2–3 oz.-doses) benefits those conditions once referred to by wise old docs as "biliousness." Frontal headaches, a bad taste in the morning, with a coated, gruff tongue (and coated, gruff personality) is what the term means, usually in the type of person that has a hankering for lots of fat, poor quality meat, and hydrogenated, crank-case grease. For some of us, eating lots of sugar or refined carbohydrates results in the desire for more of the same. For others, lots of rancid or oxidized fats can bring out the desire for and metabolic reliance on more of the same. This is subtle stuff at work, and a seldom examined phenomenon in standard practice medicine, but a reality nonetheless.

Some of the *Artemisias,* including this one, have been shown to help decrease the ill effects of lipid peroxides (rancid fats) on the liver. In dealing with the American diet and our strange reliance on processed foods, you must remember that fat and oil preservatives and antioxidants don't change the existing rancidity, they just prevent more from happening and your finely tuned nose from detecting what has already happened. That bag of questionable donuts may cause short-term pleasure, a few hours of malaise, and metabolic white-noise . . . and the desire for another bag. Try a cup of the evil-tasting cold infusion before retiring and often the lipid addiction will fade somewhat the next day, and your vague greenies will vanish. See Formula #3 (pg. 137) for a tea that combines well for this condition with tinctures of Algerita (or Oregon Grape) and Milk Thistle.

The hot tea induces sweating, expectoration, and a thorough cleaning out of the sinuses when you start to feel those cold and flu symptoms. The oils are somewhat anti-inflammatory and anesthetic, enough so that a strong tea can be applied to sprains and bruises, or the dried herb moistened and applied as a poultice.

OTHER USES: The young shoots, one to one and one-half feet tall, in April or early May, can be made into moxa, and used when strong heat is needed for very cold conditions in moxabustion.

CONTRAINDICATIONS: Because of its stimulation of the uterine mucosa, it should not be used in pregnancy.

CAMPHOR WEED

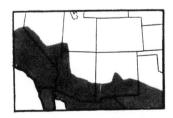

Heterotheca subaxillaris or *H. grandiflora*   Compositae
OTHER NAMES: *H. latifolia, H. scabra,* Mexican Arnica, False Arnica, Telegraph Plant

APPEARANCE: These are two interbreeding species, and both are big annuals or biennials, from two to six feet tall. They have large, alternate, stem-clasping leaves and a dozen or more medium-sized flowers that look a little like Dandelions and a little like daisies. The stems are thick and gummy, as are the leaves. Various environmental detritis usually cling to the sticky foliage in the same manner as Yerba Santa, native Tobacco, and Mullein. The whole plant (especially *H. subaxillaris*) has a strong piney-camphory smell. The eastern species can bloom all year long; the western one blooms almost entirely in the fall.

HABITAT: *H. subaxillaris* is found along roadsides and waste places throughout the Sunbelt, from the Atlantic Coast to, by now, coastal California; it is most abundant from western Arizona to West Texas. *H. grandiflora* is native to southern and central California but is making its way in the *other* direction, now found as far east as western New Mexico.

CONSTITUENTS: At least 7 sesquiterpenes, 2 dynenes, cadalene, and quercetin.

COLLECTING: The whole flowering plant, dried Method A. Try to gather clean plants, but if they are unavoidably dirty, wash them in cool water, squeeze them out by hand, and shake the foliage apart so the sticky leaves don't cling together and spoil while drying.

STABILITY: For up to a year, or as long as the characteristic scent is strong.

PREPARATION: Although traditional Mexican usage calls for the flowering heads, I have found the whole plant to be useful. Grind up the dried herb and tincture either in 70% isopropyl alcohol (if you don't mind the low-class smell) or 1:5 in 60% alcohol, Method B. Use Method A for the salve.

MEDICINAL USES: A tea of the plant or the diluted tincture is antiseptic and antifungal and may be used whenever there is a need for a cleansing wash prior to bandaging or dressing a moderate abrasion, scrape, or cut.

More especially, the tincture or salve is a first-rate liniment or ointment for any sprains, dislocations, or hyperextensions. It doesn't work quite as well for treating arthritis as True Arnica (see *MPOMW*), but it works nonetheless. As True Arnica doesn't grow below 9,000 feet much, nor south of the southern Rockies, it's nice to have this effective substitute, thanks to Mexican usage. The trick with the tincture or salve is to use it when there is pain on movement. (It doesn't hurt you to sit down and watch television or read this book; it hurts when you get up to turn it off or go outside to look for the plant, book in hand, a knowing glint in your eye.)

CANADIAN FLEABANE

Conyza (Erigeron) canadense   Compositae

OTHER NAMES: Horseweed

APPEARANCE: A tall annual or biennial, with slender, erect stems, varying from a few inches to seven or eight feet in height. It branches toward the top as it matures, bearing many small flowering heads. The stems are hairy, although local populations and growing conditions will vary the fuzziness. The leaves are long, thin, and strap-shaped, from one to four inches in length; the lower ones are often slightly notched, the upper ones are entire. The flowers are the ultimate in nondescript, semi-yellowish, vague drabness. This plant is so common in some localities that it is hard to focus on it as a distinct identity. Along with tumbleweeds, dandelions, and pigweed, it is the plant equivalent of grey noise.

HABITAT: Native to the eastern United States (not all our weeds are Eurasian adventurers), it is widespread along roadsides, ditch banks, cultivated fields, and waste places, usually forming widespread communities. The best plants are tall and light green; those around the raggedy edges are small and scruffy. Like Yellow Dock, Añil del Muerto, and Puncture Vine, the aesthetics of a gathering trip for Canadian Fleabane are (yawn) minimal. The aesthetics of the medicinal uses are rewarding enough.

CONSTITUENTS: Limonene, dipentene, proprionic and gallic acids, terpeneol, and methyl caprate derivatives.

COLLECTING AND PREPARATION: Gather the whole plant, roots and all, when in early flower, Method A. The essential oils are the major therapeutics, and the plant needs to be stored in whole form and chopped into tea as it is needed. For most conditions, ¼ ounce of the herb in a Standard Infusion per day is best, although for moist coughs and runny nose the simple aromatic tea, somewhat pleasant tasting, is appropriate at any dosage and weight.

MEDICINAL USES: The infusion is an excellent treatment for diarrhea, profuse sweating, and the excessive production of slightly irritating urine, as well as most allergic-type hay fever and excess bronchial secretions. The trademark for Canadian Fleabane is congestion, not inflammation; not acute (hot) but subacute and chronic (cold). For hemorrhoids that linger for days, never get especially inflamed or painful but always have a dull ache and "presence," may bleed a little but are never sharp and painful, Canadian Fleabane is one of the best treatments. The same is true for diverticulitis of a similar type. For ulcerative colitis, with recurring episodes of diarrhea that are not yet severe enough to be treated well medically, but with a well-established medical diagnosis to rule out more serious pathology, Canadian Fleabane is the answer, taken ¼ ounce of herb in infusion a day for at least a week.

Irritable bowel syndrome (IBS) responds well to this little weed, especially when the cholinergic phase is well established and the diarrhea episodes are painful and aching.

Although Shepherd's Purse is a better herbal treatment for short term, acute mucous membrane bleeding, Canadian Fleabane works very well for bleeding from congested, edemic, boggy tissues. Combine with Milk Thistle, Prickly Pear Flowers, and Shepherd's Purse for such uses. See Formula #3 (pg. 137).

The leaves (not flowers) can be finely powdered and snuffed into the nostrils for acute hay fever with profuse nasal discharge and sore, itchy eyes. An old Navajo treatment.

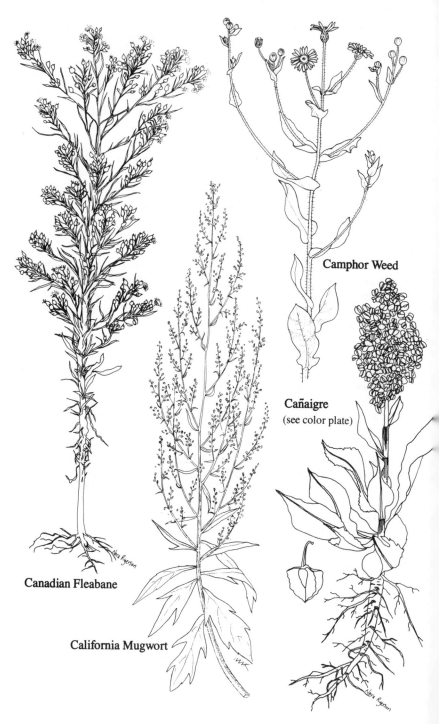

Camphor Weed

Cañaigre
(see color plate)

Canadian Fleabane

California Mugwort

CAÑAIGRE

Rumex hymenosepalus   Polygonaceae

OTHER NAMES: Red Dock, Pie Plant, Tanner's Dock, Wild Rhubarb

APPEARANCE: Cañaigre has thick, somewhat succulent leaves, one-half to one and one-half feet long, lance or spade shaped, wavy, with a pronounced central vein, and sour to the taste. Leaves are mostly basal, trailing along the ground with occasional alternate shorter leaves clasping the single stout stem, which grows from one to three feet tall. The top half of the stem bears many inconspicuous green flowers in the spring that mature into a thick, showy cluster of pink-red seed pods, three winged and somewhat heart shaped. The most definite identification comes from Cañaigre's underground cluster of yamlike tubers, from two to over a dozen. They are dark reddish brown with an inner flesh ranging from light orange to dark, rust red, intensely astringent to the taste.

HABITAT: Surprisingly varied. It grows in luxuriance from the deep arroyos of the Colorado Desert to the foothills of the Sangre de Cristo Mountains of New Mexico, from sea level to 8,000 feet, although most frequently in dry, sandy soil from 3,000 to 6,000 feet. The best plants for medicinal use tend to be found in and around sandy washes of the high deserts, 3,000 to 5,000 feet. Best gathered in early spring—March and early April—before full flower. After May, they disappear into the general environment, dying back until the following spring.

CONSTITUENTS: Dry roots are 20% to 30% tannins, with potassium oxalate crystals, emodin, chrysophanic acid, and 1% to 2% anthroquinones, including physcion (leaf and root), leucocyanidine, leucodelphinidine and leucopelargonidine.

COLLECTING AND PREPARATION: The leaves as a fresh Method A tincture or dried for a Method B salve; the roots dried. They are easier to dig if growing in elevated ground and sandy soil, difficult or impossible if found in gravelly basin sediment. Although very slow to rot, the tubers are best cut into one-quarter-inch cross-sections; a whole dried tuber can be a heavy, purposeless thing. The slices should be spread on paper and sun dried. Dry the leaves normally, Method A.

STABILITY: The tuber whole, sliced, ground, is stable for years. The ground leaf for six to eight months.

MEDICINAL USES: Like other *Rumex* species, such as Yellow Dock (see *MPOMW*), the roots are biochemically complex, but Cañaigre root is so high in tannins as to serve little other function than as an astringent. If you are sunburned to any degree in a desert area where both Cañaigre and Prickly Pear grow, a not unlikely combination, the root can be grated fresh onto the burned skin, allowed to dry, and a filet of Prickly Pear placed over the same area, or its juice rubbed in. The roots are boiled for an astringent and hemostatic wash for cuts or scrapes, or used as a mouthwash or gargle for ulcerations of the gums and oral mucosa.

A recent study by a team of allergists has shown that the leaves, fresh or dried, have probable antihistamine, anticholinergic, and antibradykinin effects; that is, they help to modify and depress local inflammations caused by hives, contact dermatitis, chafing, or simple friction. I have found that the alcohol-intermediate salve made from the dried leaf, or a tincture, either of the fresh leaf in pure alcohol or in three parts vinegar, are all effective in these cases. It turns out the Navajos have used the leaf for this purpose for centuries, especially for sores and inflammations on the face, mouth, and neck.

California Indians have used the dried seeds with much success in controlling the magnitude and pain of diarrhea and dysentery, a handful boiled up in some water and drunk when cooled down to body temperature, or simply steeped in room temperature water overnight.

A final note: *Panax quinquefolium* (American Ginseng) and Cañaigre have nothing in common, pharmacologically or botanically—no similarities in function, habitat, appearance, or taste. Cañaigre is a robust, common, and wide-spread plant, whereas wild American Ginseng is so rare as to approach over $200 a pound wholesale. Since there are no legal standards to the "Ginseng" name, several companies in the Southwest are marketing Cañaigre, grossly overpriced, as "Wild Red Desert Ginseng," "Wild American Red Ginseng," and other misleading names. The reddish color acquired in the curing of several grades of Oriental Ginseng (*Panax ginseng*) and the lack of knowledge on the parts of some health food stores and the public enable this deception to continue.

OTHER USES: The leaves can be substituted for rhubarb in making rhubarb pie.

CENIZO

Leucophyllum
frutescens   Scrophulariaceae
OTHER NAMES: *L. texana*, Purple Sage,
Palo Cenizo

APPEARANCE: Cenizo is a handsome spreading shrub with many grey, silvery, simple leaves, crowding the green-tinged stems. The lower branches are reddish brown and tend to get shredded bark. The silver leaves persist more or less as an evergreen. The flowers are large and showy, up to an inch across, reddish purple with dark spots on the throat; they bloom from the leaf axils and terminal stems. In good locations the bushes may reach six to eight feet in height, and are beautiful when blooming in late summer and early fall, the bushes covered with silvery foliage and purple flowers. They are often cultivated as ornamentals.

HABITAT: Texas, from Brewster and Pecos counties east to Travis County and south to the Rio Grande. *Leucophyllum minus*, a smaller bush with the same use, grows west of the Pecos River, through to eastern and southeastern New Mexico. It is smaller, with smaller leaves and usually fewer flowers, but otherwise has the same handsome demeanor.

CONSTITUENTS: Cyanidine-3-rutinoside and an unnamed alkaloid.

COLLECTING AND PREPARATION: The branches gathered when in flower, dried Method A, and then stripped of the flowers and leaves; discard stems. Like any other simple tea, it is brewed as a simple hot infusion, a teaspoon or rounded teaspoon of the herb in a cup of boiling water.

MEDICINAL USES: Widely drunk in Texas and Chihuahua by Mexicans and Anglo ranchers alike, it is your basic cold and flu tea, straight or with a little Peppermint or Pennyroyal (see *MPOMW*). The effect is straightforward and effective; it induces sweating, helps to break up the fever, and is sedative enough to help allow sleeping.

Just as important, it is one of our good-tasting native teas. Along with Cota, Mormon Tea (see *MPOMW*), and Yerba de Alonso Garcia, it is a cheerful, sensible, drink-for-the-pleasure tea. Part of the gentility of Cenizo may be the small pleasure of having the jar filled with silver and purple herb that you dip into periodically.

CHAPARRAL

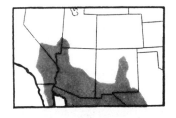

Larrea tridentata   Zygophyllaceae

OTHER NAMES: *L. divaricata, L. glutinosa, L. mexicana, Covillea* spp., Creosote Bush, Greasewood, Gobernadora, Hediondilla

APPEARANCE: Chaparral plants are leafy, reptilian, many-branched bushes, up to twelve feet tall, but usually chest- or head-high. Under moist conditions the foliage is a rich, greasy, yellow green, but waxy olive-drab during drought or freezing and dead brown in extreme drought. The leaves are small and curled and the same color as the photosynthesizing smaller branches; the larger branches and trunks are reddish brown to black. The flowers are yellow and cute. They cover bushes after a good rain any time of the year and mature into fuzzy round capsules. Real creosote is a derivative of pine oil or coal, but it's the only other thing that smells quite like Chaparral.

HABITAT: Every square mile below 4,000 feet, from eastern San Diego County all the way to central Texas, is likely to have some of the plants growing on it. It can grow as high as 5,500 feet (south of Albuquerque), but below 3,000 feet, our deserts are owned by Chaparral. The Mexican name of *Gobernadora* (the Governess) describes its ability to form a monoculture in many deserts; the New Mexican name of *Hediondilla* (Little Stinker) needs no explanation.

Our common name of Chaparral makes botanists and range-management types grind their teeth in frustration; for them, Chaparral is a name for a whole biosphere of life ecology, like arctic tundra or oak woodlands. Nonetheless, it is our most common name for the plant as a medicine.

CONSTITUENTS: 18 distinct flavone and flavonol aglycones, a dihydroflavonol, larreic acid, two guaiuretic acid lignins including (most important) nordi-hydroguaiuretic acid (NDGA), and several quercetin bioflavonoids. NDGA content ranges from 1-1½% of the dry plant.

COLLECTING: Good, strong, well-foliaged plants are best, whenever gathered; dried, Method A. Strip the leaves, flowers, seeds, and small twigs off the branches and discard the woody stems. For external, antimicrobial use, older plants work better; internally, leafy, bright green youngsters are preferred. The difference is only minor, so gather what's best for where you are.

STABILITY: The dried plant is usually stable for up to two years.

PREPARATION: The tincture, Method B, 1:5, 75% alcohol; the salve, Method B; the powder in #00 capsules. Now comes the problem: "Now that I have some of this stuff, how do I grind it up! It just sets in the bottom of my blender after turning into a creepy green goo that started to smoke from the friction!" You freeze it. If you have a blender, fill it halfway with the herb and put the whole container in the freezer, remove after an hour, grind for a few moments, strain the powder out, put the coarse pieces back into the container, return it to the freezer, and do the whole thing again. Put the remaining coarse pieces back with the whole herb for the next time. If you are using a hand grinder, freeze the herb and the grinder first. A friendly hint: don't try to powder Chaparral with a stone-grinding head (under any circumstances!). Four years later that freshly milled, hard winter high-pro wheat you ground into flour will still taste like the Chaparral you ground four years earlier.

MEDICINAL USES: What Chaparral does comes from its ability to inhibit aerobic combustion in the mitochondria of cells. In the desert, the oils leeched out into the surrounding soils inhibit seeds from burning up their sugars; the seeds can't sprout unless the oils are washed away by heavy rains. (Chaparral even inhibits its own combustion and burns with such a low flame that there is a plant in Imperial County, California, that has been carbon-dated to be over 13,000 years old . . . now, that's slow!)

When applied to the skin as a tea, tincture, or salve, Chaparral slows down the rate of bacterial growth and kills it with its antimicrobial activity. This makes for a very useful first aid or long-term dressing for skin abrasions and injuries, for man or beast (not to sound sexist or species-ist).

Speaking practically, a tea made from Chaparral is nearly undrinkable. I would go so far as to say that if you *like* its taste you may need professional counseling. This limits "patient compliance" and probable consumption to the tincture or capsule. Internally, it has a strong and beneficial effect upon impaired liver metabolism. If you have dry skin, brittle hair and nails, and cracks in the feet or hands, but you have trouble digesting or even absorbing decent dietary oils, then a regimen of Chaparral will help; 30 drops of the tincture or 2 capsules before retiring. Some clinical trials have shown that it aids individuals who also manufacture poor-quality blood lipids. In blood chemistry workups, this manifests as elevated LDLs (Low Density Lipids) and even (for the heavy drinker) VLDLs (Very Low Density Lipids). These transport fats really wreck the blood vessels, leading to arteriosclerosis and hardening of the arteries.

Chaparral combines very well with Milk Thistle and Puncture Vine to control the problem.

In the recent field of radical oxygen chemistry, Chaparral has been shown to contain compounds that inhibit the damage to the liver and lungs from the elevation of these free radicals, and, with Milk Thistle or alone, it decreases the ravages of rancid, dietary lipid peroxides, especially on the hepatocytes of the liver. This gives it useful application in the treatment of joint pain, allergies, autoimmunity diseases, and even Premenstrual Syndrome. As for the often mentioned effects it has on cancer, further tests over the years (yes, they sometimes take this stuff seriously) have shown that Chaparral can both inhibit and stimulate the growth of cancer cells, so forget that. For the other conditions, the small daily doses recommended are the best approach.

OTHER USES: Chaparral's antioxidant abilities make it useful to prevent rancidity of vegetable oils . . . if you aren't going to cook with them. Four tablespoons in a quart of oil, steeped for a week and strained, gives you stable oils to use for massage, ointments, or salves, even as a mild sunscreen. Its ability to inhibit plant growth would seem to make it a perfect way to cut back your fast-growing junkers, like Bermuda Grass and the like, since the effects wash out when needed. If you figure out how to do it, and find out that it does or doesn't work, please write and tell me.

CHAPARRO AMARGOSA

Castela emoryi and *C. texana*   Simarubaceae

OTHER NAMES: *Holacantha emoryi, H. stewardii,* Crucifixion Thorn, Corona de Cristo, Amargosa

APPEARANCE: This bush has no leaves, unless you wish to count a few little scales here and there. The branches are stout, thick, rigid, and distressingly sharp; that is all there is . . . thick, rigid, and sharp branches, intermeshed with each other in bizarre patterns, crisscrossing and crossing back again, like the olive green insides of an Iron Maiden left over from Torquemada's Inquisition dungeon. The plants perform their photosynthesis in the stems and periodically decide to reproduce, especially if there is another plant of the opposite sex nearby. Some are female, some are male . . . stamens only a pistil could love, I guess. I have seen grumpy little stands of a dozen males, miles from any females, and later seen a ten-foot-tall female, covered in several years' growth of little brown seed clusters, miles from a male. The young stems have a fine dusty bloom to them, the older stems get covered in streaked brown grey bark and call themselves trunks; the spikes that arm each stem are either darker or lighter than the greenish stems. A weird plant. It can be confused with Allthorn (*Koeberlinia spinosa*), but the latter plant grows in somewhat higher altitudes, has shiny black berries instead of clusters of reddish brown seeds, separating

29

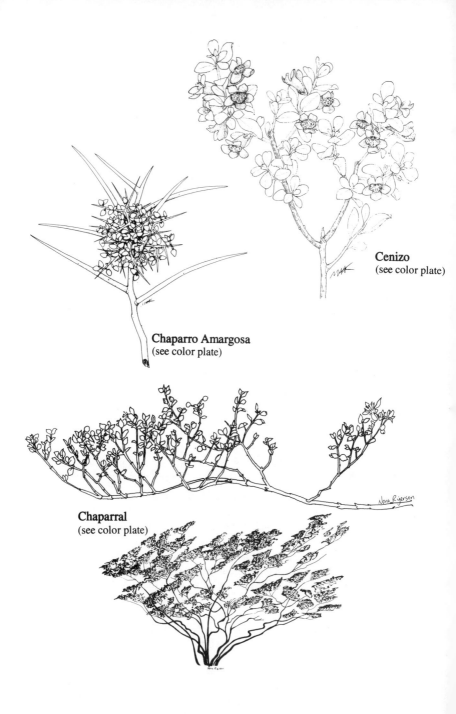

Cenizo
(see color plate)

Chaparro Amargosa
(see color plate)

Nora Ryerson

Chaparral
(see color plate)

into five or ten drupes like a Star Anise, and the flowers are greenish white, while those of Chaparro Amargosa are reddish brown. Allthorn is dark green, Chaparro Amargosa is bluish grey-green.

HABITAT: From 500 to 2,000 feet in the driest, hottest, nastiest parts of our deserts, especially in coarse sand or fine gravel soil, from Daggett, California, to Florence and Eloy, Arizona, south into Sonora and Baja California. *Castela texana*, our other Chaparro Amargosa, a somewhat smaller and broader plant, can be found from south of El Paso, through the Big Bend of the Rio Grande and down toward the Gulf of Mexico, usually in sand dunes. This one is rumored to have little bitty leaves on occasion . . . I'll believe it when I see it.

CONSTITUENTS: Chaparrin, Chaparrol, isochaparrol, neochaparrol, chaparrinone, glaucarubol, chaparolide, castelanolide, ellargic acid, and additional chaparrol lactones.

COLLECTING: Very, very carefully. Cut off a single stem down where it starts to get barky, unravel it slowly from the rest of the mess, carry it back facing away from you. Chop it up into little sections while still fresh, place the pieces in a brown paper bag to dry, and breathe easier. The hard part is done.

STABILITY: Whether chopped up fresh or stored in a whole branch, it will be stable for years.

PREPARATION: The tincture is the most practical, Method B, 1:5, 40 or 50% alcohol; it is water soluble but the Spanish name means Bitter Bush and they aren't lying. Gulping a dropper or two with a chaser is easier to handle than sipping a cup of yellow green nastiness, especially if you have the runs and seriously impaired patience. The tincture, 25-50 drops every several hours.

MEDICINAL USES: Like many of its relatives, such as Quassia and Cascara Amarga, Chaparro Amargosa is a very active inhibitor of intestinal protozoa. Amebic dysentery (from *Entamoeba histolytica*) and giardiasis (from *Giardia lamblia*) respond well to the plant. It will predictably prevent or limit the magnitude of infections *before* they occur, used as a preventative when traveling long distances for vacations and such. This does not mean "south of the border" particularly, since going long distances in any direction from where you are living means confronting a variety of soil and food microorganisms that are almost, but not quite, the same as your gut is used to; the result: traveler's diarrhea.

Giardia is a little flagellate protozoan that looks like a character out of an old Disney cartoon and can make you feel like something out of an old Lon Chaney movie. It is found in pristine, beautiful mountain streams, trickling down from beer-commercial snowfields . . . not to mention in the contents of the occasional tostada purchased on the plaza somewhere in Sinaloa.

To inhibit any microorganisms in the water you are drinking, squirt into it half a dropper of the tincture, wait a minute, and drink. If you can't stand the taste, squirt it in your mouth, *then* drink the water. If it's the food you are worried about, do four or five squirts a day and forget about it. Chaparro Amargosa is useful in herbal treatments to reorganize the intestines after antibiotic therapy or lingering gastroenteritis, with or without a Candida suprainfection. Combine it with Desert Willow, Echinacea, Algerita, Yellow Dock (see *MFOMW*), Silk Tassel, or Arizona Walnut.

OTHER USES: Chaparro Amargosa can also be used to make your own poisoned flypaper. See Formula #5 (pg. 137). If the flies don't stay stuck but fly away, they will die later (giggle) of chaparrol lactone poisoning.

CHICKWEED

Stellaria media . Caryophyllaceae

OTHER NAMES: Starweed

APPEARANCE: If you have a moist, shady place in your garden, you probably have some Chickweed growing there also. It is a little, lime green ground cover and matting plant, with foot-long jointed stems and opposite oval leaves, mostly clasping the stem, sometimes with short stalks. The flowers are snow white like little stars, five petaled, but with each one so deeply cleft that it seems to be ten-petaled. The seeds are green capsules, and as summer wears on, the stems, capsules, and the lower leaves start to turn papery tan. It has a slightly succulent, salty taste and, when gathered into bunches to dry, a mild, brackish scent.

HABITAT: Moist, shady places under larger plants and shrubs or, in early spring, along sidewalks, steps, and the edges of vacant lots. It is better established in older gardens and alleys of the inner city, less likely in new subdivisions. Although an herb found worldwide, it is a stronger medicine in the heat of our warmer climes.

CONSTITUENTS: 6, 8-di-C-galactosylapigenin and (2)-C-glycosylflavonoid.

COLLECTING AND PREPARATION: Chickweed is best fresh or as a fresh tincture, Method A. The dried plant is serviceable for up to six months, gathered in mats, garbled to remove twigs and detritis, and dried in flats. The salve, from recently dried herb, Method A.

MEDICINAL USES: The fresh plant (mashed for a poultice), the fresh plant juice (washed, shaken, juiced), the tincture, or the salve are all useful topically for swellings. It is most effective for fingers, hands, and foot swellings from sprains, arthritis, gout, or pseudo-gouty conditions; deeper pain or contusions are too far into the body for Chickweed to help. Its effects on smaller distal joints is distinct and beneficial.

The dry plant is a useful diuretic for premenstrual water retention, or retention in athletes using steroids or eating a pound of beef a day to put on (some sort of) weight, with over-acidic urine and concurrent sodium retention from anabolic-excess stress. As for the use of Chickweed to reduce weight, as touted by some nutritionists and pyramid-sales "herbalists," I have little doubt that it helps birds.

OTHER USES: If you have canaries or parakeets, give your pets some of these greens fresh . . . they love it.

CHIMAJA

Cymopterus fendleri   Umbelliferae
OTHER NAMES: Biscuit Root, Chimaya, Chimaha

APPEARANCE: This little spring beauty looks like a patch of Italian Parsley with big yellow Parsnip flowers stuck in the middle. The whole plant radiates out from a central root and is much broader than tall. In March and early April the roots put out one or two little Parsley leaves and a squat umbel of yellow flowers. As it grows there are additional leaves added to the rosette and maybe another flower umbel or two. By the end of May the whole plant has disappeared, reverting back to the long, tasty tuber, waiting until the next spring. Leaves are divided several times into little wedge-shaped sections, dark shiny green and curled up at the edges. Many other similar plants, such as other *Cymopterii, Pseudocymopterus,* and some *Lomatiums*, resemble Chimaja, though the leaves of these imitators are dull or fuzzy, the flowers are not yellow, or the root is not tuberous. In any respect, none of these low-growing munchkin-parsleys have Chimaja's lovely taste.

HABITAT: The center for Chimaja is the San Juan Basin and the high deserts of central and northern New Mexico, always growing in sandy, gravelly, dry mesas and low-slung foothills, from 4,500 to 6,500 feet, perhaps a bit lower in West Texas and Chihuahua. It grows in the Mogollon highlands of Arizona, along the flat hills of the Little Salt River to the foothills of Arizona's White Mountains, the barren mesas of eastern New Mexico, southwestern Colorado, etc. It is associated with *Ephedra torreyana* (one of the Mormon Teas), Maravilla (see *MPOMW*), and open Juniper/Piñon hills, always in colonies.

PREPARATION AND MEDICINAL USES: The leaves and seeds, like those of its high-mountain relative, Osha (see *MPOMW*), are put in bottles of brandy and corn liquor, to help take away the stomach pain that you get from drinking too much . . . brandy and corn liquor. In the old days, restaurants in Albuquerque and Belen offered the seeds as an after-dinner digestive aid. If what I have heard about the carne adovada and green chile stews in the old days is true, it may have proved a life-saving measure. The root can be roasted, split, and placed over abcesses and boils, to focus the inflammation and bring it to a head. The leaves are used by the Hopi and Navajos to stimulate slow-starting, crampy menstruation and to help urinary infections.

FOOD USES: The leaves taste exactly like Cilantro Leaves; however, unlike Cilantro, they can be dried and still retain their flavor. Chimaja is an impeccable spice (leaves or seeds) for mixed-media dishes, such as vegetable-meat stews and barley-lamb soup and the like. The seeds are a strong, spicy cross between cumin and celery, with the best uses of both. I once had some exquisite German potato salad, hot and steaming, flavored with Chimaja seeds. The roots, washed and peeled, then steamed, taste like parsnips and turnips only wished they could; freeze the cooked tubers for later chopping into wok-fried veggie dishes or dice and brown them in butter for soups.

CLIFFROSE

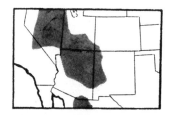

Cowania mexicana   Rosaceae
OTHER NAMES: *C. stansburiana*, Quinine Bush

APPEARANCE: This is a medium to tall shrub, with open and erratic branches having shreddy, grey bark on older, thicker trunk stems, and shiny, reddish brown bark on the younger branches and twigs. These are profusely covered with little palmate or semi-compound dark green leaves that are crinkly, leathery, and almost succulent. The bushes seldom reach their full potential size since everything that is on hooves eats the middle and lower leaves, leaving the plant open and irregularly fan-shaped. In May the plants burst forth in hundreds of cream-colored little roses having a strong, almost narcotic orange-blossom scent; they often bloom on into the fall if there are enough summer rains. From five to ten in a bunch, the seeds are plumed, about two inches long, and resemble those of their close relative, Apache Plume.

HABITAT: Cliffrose is common in the Great Basin, making it southwards to the foothills of the southern Arizona and northern Sonora mountains and the edges of the Mogollon plateau; from 3,500 to 8,000 feet. Look for it, oddly enough, on cliffs, hillsides, and limestone rubble. In the spring, just follow your nose up the hills.

COLLECTING: The buds and new blooms deprived of the green calyx and carefully dried, Method B. The leafing brown branches, bundled and dried.

MEDICINAL USES: The chopped and boiled stems and leaves make a soothing, if bitter, cough suppressant, especially useful in the early stages of a chest cold with a hot, drawn sensation and a dry, irritable throat. Drink it slowly, gargle a bit, and swallow. It also will induce sweating and the softening resecretion of mucus.

The same tea aids backaches that cause a vague need to urinate, or defecate, or sleep, or something . . . not sure what, exactly. The kind of backaches that truck drivers are famous for getting.

The flowers and buds, deprived of their bitter green calyx, can be added to other teas for their added delicate fragrance. Too many in a pot of herb tea and the delicacy becomes bitter and somewhat unpleasant, so use with a light touch.

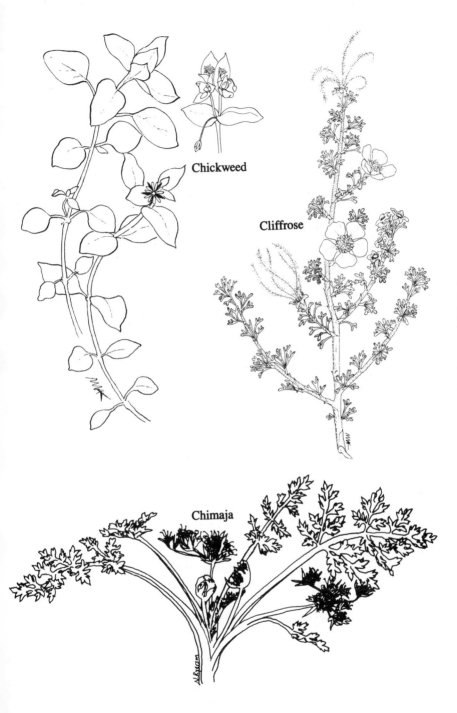

Chickweed

Cliffrose

Chimaja

35

CONDALIA

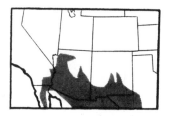

*Condalia lycioides (Zizyphus obtusifolia), C.
spathulata, C. mexicana*   Rhamnaceae
OTHER   NAMES: Lotebush,   Desert
Lotebush, Lotewood, White Crucillo,
Greythorn, Abrojo, Tecomblate

APPEARANCE: Although the genus of several of these plants has been changed,
(*Condalia lycioides* and the closely related *C. obtusifolia* are now *Zizyphus*), the
appearance is still the same; big, spiny, impenetrable shrubs, with little dark
leaves, alternate and delicately veined, all up and down the spines and stems.
Flowers appear in the spring, little cream to light greenish white clusters in the
leaf axils. These mature into blue, brown, or blue-black, fleshy berries, mealy
but tasty. The exception to tasty is the Bitter Condalia (*C. globosa*) of the Colorado Desert foothills . . . nasty. All the Condalias tend to interbreed or change
form and foliage drastically in different areas—like their relative, Red Root (see
MPOMW)—and every text calls these lovely little bush-trees, with their nasty-
tipped right angled stems, by different specifics and even genera. Lamb (*Woody
Plants of the Southwest*) describes the problem well: "The form seen on the
Pecos drainage . . . is a low, leafy, dark-colored bush; from Tucson westward
[it is] a tall open-growing shrub, very spiny, light grey in color, and almost leafless." Some of the Condalia types, nearly naked-stemmed in the summer, are so
densely twigged, branched, and spiked that they look like grey Junipers; a little
rain and they become dark green, covered with their neat little leaves.

HABITAT: The dry belt between Juniper/Piñon/Ponderosa and the low, flat desert. Condalia likes arroyos, mesa rims, bajadas, and gnarly, washed-away alluvial fans. From the Grand Canyon to the Anza-Borrego, from the Superstitions
to the Sacramentos. If you stand in a valley, look up to the dark trees at the top;
you will see vast seas of round dark shrub-trees mottling the middle sections,
like stippling in a line drawing. Many of these will be Condalias.

COLLECTING AND PREPARATION: In the eroded environment where the Condalias
often grow you will frequently find their large dark-barked roots snaking
nakedly over rocks and rubble, exposed to the air from some cloudburst's
ferocity years ago. Saw or snip off the root close to the tree and wedge out the
distal section. The plant is hearty and disease-resistant, and this won't harm it.
This fresh root, chopped, peeled, or grated, is used for soap. The dried chopped
root is boiled for medicine.

MEDICINAL USES: The boiled fresh root makes a sudsy soap, used by the
Apaches for washing their hair. It makes a useful treatment, used regularly, for
scalp sores, tinea capitatis, and seborrhea. For increased soapy effect, add some
Soapberries or fresh, grated Yucca Root (see *MPOMW*).

The dried root tea is astringent and antiseptic with a mild but distinct expectorant effect. Comanches and Apaches used the strong decoction to bathe sores
from scraping, abraded skin, friction blisters, and windburn; they also used it to

bathe the backs and flanks of injured horses. It will take away pain and at least some of the inflammation while cleaning and disinfecting the wounds. A little of the tea as an eyewash helps remove the inflammation that occurs in conjunctivitis and sore eyelids. Snuff up a teaspoon of the tea from your palm into the sinuses if you have rhinitis or hay fever. The decoction loosens up tough bronchial mucus in the waning days of bronchitis.

Pimas used the thorns to prick the skin around sore arthritic joints, or sprains and contusions after they were cooled down a bit. Acupuncturists use Plum Blossom needles and moxa for the same purposes, and it wasn't too many years ago that physicians achieved a similar therapeutic counter-irritation by inducing little blisters around the joints with tincture of Cantharides (Spanish Fly), Mustard Oil, or little, hand-cranked friction burners. It's a little drastic to us these days, but a really sore knee responds well to these treatments.

FOOD USES: Except for the previously mentioned Bitter Condalia, the fruits are tasty; dry well for adding to granola or applesauce. Pull off one end of the fresh berries, squeeze out the seed, and put the squished-in fruit on flats to dry. The seeds can be ground in a grain mill (after dry) for adding to oatmeal or breads; they are nutritious and contain up to 25% linoleic acid. The seeds or whole fruit would be good to add to feed for cattle or goats.

COTTON ROOT

Gossypium thurberi (native) and *G. herbaceum* (cultivated)   Malvaceae
OTHER NAMES: *Thurberia thespesioides,* Algodoncillo, Desert Cotton

APPEARANCE: This is a handsome, sparse shrub, averaging four to eight feet in height, with gracefully tapering three-lobed palmate leaves, three to four inches long. They have further partial divisions, but three-lobed is predominant. They all turn maple-leaf red in the late fall; in fact, reddish brown is a major part of the plant's coloration. The flowers bloom in July and August and are two or three inches across and Hollyhock-like. They have many colors, but a light sulphur yellow is most common and white and lavender pink less so. The cotton bolls are little black balls, one-half inch in diameter, that split into three segments when ripening in the late fall. The cotton fibers are much too short to be useful; the seeds are nutritious when roasted.

HABITAT: The native Cotton grows between 2,500 and 5,000 feet on gravelly and rocky hillsides and along the lines of high-water flooding in arroyos and small valleys. It is found only, but abundantly, in Arizona and Sonora. Cultivated Cotton is grown in much of our area, usually in a cesspool of agribusiness chemicals that can linger in the root. You can gather this plant with safety in some areas, where the chemicals are unnecessary, if you follow these guidelines: *Defoliants* are not used if it is grown high enough and cold enough that the first major frost kills back the leaves. *Herbicides* are not used if there are little dead annuals growing up between the rows of Cotton, and it is grown, therefore, in a cold winter area where one good turning of the soil is enough to keep out perennial weeds. *Pesticides* are not used where the winters are harsh enough that boll weevils cannot survive.

Cotton is mainly a tropical perennial that dies from frost; it is cultivated as an annual, so the roots are never as strong as our native Cotton (frost-tolerant), which may grow for a decade or more. Still, you get what you can; crops in West Texas, New Mexico, and eastern Arizona usually need the least care and are usually the least chemically abused.

COLLECTING: The bark of the native species, fall to early spring; cultivated plants after the leaves have frozen back and the cotton bolls are harvested; after harvesting, it is unlikely that the farmer cares about the naked stems, but always ask permission. Cut the roots about an inch above the demarcation between the smooth, brown plant stem and the rougher, reddish brown root bark; wash them and pull off the bark from the core in strips.

CONSTITUENTS: Acetovanillin, phenolic acid, betain, phytosterols, and a yellow chromogen; none of these seem to account for its pharmacologic effects.

PREPARATION AND STABILITY: Best, fresh root bark tincture, Method A; Good, recently dried root bark tincture, Method B, 1:5, 50% alcohol; Adequate, recent dried root bark boiled up for tea. It can be kept fairly active for up to a year if it is dried, sealed well, and frozen to slow down chemical degradations.

MEDICINAL USES: The sole activity of Cotton Root bark seems to be to stimulate those tissues that respond to the pituitary hormone oxytocin. It facilitates and acts synergistically with the hormone; without some oxytocin present in the bloodstream, the herb has no effect. It therefore is used to increase the tone and contractability of the uterus, seminal vesicals, and the myoendothelial tissues of the breast.

In childbirth, it is a valuable aid for home deliveries, where pitocin (synthetic oxytocin) can be inappropriate, but a little help wouldn't hurt. It should be used in ½–1 teaspoon doses of the tincture in a little warm water, or ¼ ounce of the bark in tea, in the first stage of labor after there has begun some obvious dilation or in the third stage to aid in placental delivery. It further helps stimulate the postpartum contractions of the nursing mother, since the neuroendocrine reflex of nipple-sucking and uterine contractions is oxytocin mediated. The let-down reflex is also increased with the herb, as it, too, is oxytocin mediated.

Some women who are nursing their babies find that after several weeks the emotional stress of parental disapproval, innate shyness, or even a covetous mate or older children can gang up to slow down and stop the milk. Such emotional stress may induce elevated adrenalin levels; adrenalin (epinephrine) inhibits many secretions, including milk. Cotton Root may help to reestablish the rhythm.

Uterine contractions in orgasm, seminal vesical contractions in ejaculation, and the erogenous effects of nipple stimulation in both sexes are all at least partially mediated by oxytocin and are usually increased by the tincture or tea. Most such problems have emotional and intellectual causes. However, if the

38

head and heart are willing but the flesh isn't, take some of the tincture or tea for a couple of days and see if that ancient engine won't turn over.

Cotton Root can substantially aid in uterine inertia, where menses is long and befuddled or starts and stops several times. The little blood vessels that constrict to separate the tissue that remains from the tissue that is discharged as menses are controlled by oxytocin. Withdrawing from birth control pills is a trying time for some women, and the first several complete cycles are better defined when small doses are taken daily. If there has been a miscarriage or a therapeutic abortion, the muscle tone of the uterus and the blood vessels is stimulated by Cotton Root as well.

For acute mastitis, where the symptoms are not yet severe enough to need medical attention, try a mixture of Cotton Root, Echinacea, and Inmortal (see *MPOMW*) tinctures. See Formula #6 (page 137).

For breast cysts that are filled with fluid, and swell and shrink with the estrus cycle, or for chronic fibrocystic breast disease, take a couple of doses a day during the acute phase, along with Red Root, as either tincture or tea. *And,* stop eating or drinking chocolate, coffee, Chinese or Japanese tea (including bancha or kukicha), cola drinks, maté, or guarana. The closely related purine-xanthine alkaloids caffeine, theophylline, and theobromine *clearly* aggravate such conditions, once they have been *clearly* diagnosed medically. Always remember that even if Medicine has dropped most of its gentler therapeutics and often has little to offer for subclinical problems except "wait-and-see," or "try this, and if it makes you sick come back and see me," or "it's probably just stress, so try some Valium," there is *no* substitute for medical diagnostics.

CYPRESS

Cupressus arizonica, C. forbesii
Cupressaceae
OTHER NAMES: *C. arizonica,* var. *glabra,*
C. glabra, Cedro, Cipres

APPEARANCE: Our most common Cypress is the Arizona species (*C. arizonica*), a medium-sized evergreen tree from thirty to forty feet tall when full grown, although a few ninety-foot giants still survive the lumber companies. The trunk is straight, with a conical Christmas tree form; the leaves are scales, the twigs square in cross-section, branching at almost right angles in all directions. The cones are usually clustered on little stalks around the circumference of the larger branches, rather than out with the foliage, as with Juniper. They are hard and woody, opening up along scale-like crevices, forming little, armored polygons with a central point. The scent of the tree is like a cross between Juniper and Pine, with a little citrus thrown in to confuse matters. Of this group, the only other species that grows in any abundance and is not endangered is the Tecate Cypress of southern California (*C. forbesii*), shorter (usually twenty feet

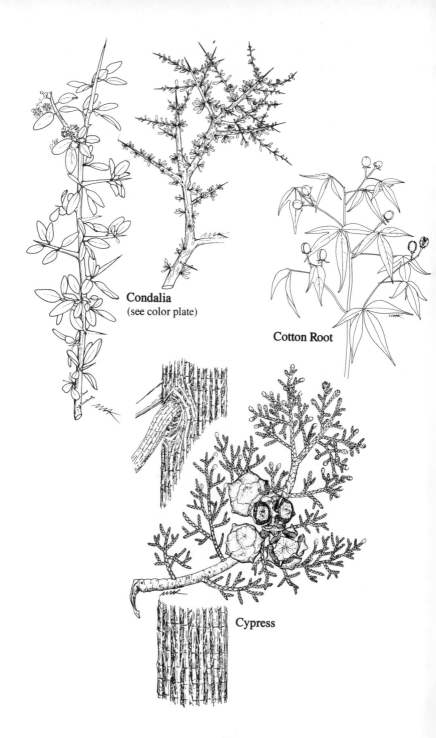

Condalia
(see color plate)

Cotton Root

Cypress

tall), with more open foliage and smaller cones.

HABITAT: Our Cypresses are remnants of ancient forests that have long disappeared; several other species in California and Nevada are found only on a few acres on a single mountain, recalling the time before the Pines and Spruces came and the ice sheets ground them down. Arizona Cypress is still with us in abundance, usually from 4,000 to 7,000 feet, from the Huachucas to the Chiricahuas, up to the Santa Catalinas and north to Oak Creek Canyon, south into northern Sonora. To the east they are found in Brewster County and in the Davis Mountains of Texas. The Tecate Cypress is found from Orange County, California, south to Baja California. Both form groves and stands in the canyons of the middle mountains and near streams where Oaks and Junipers line the canyon walls. The others should be left to their ancient vigil.

CONSTITUENTS: Cedron, thujone, furfural, d-pinene, d-camphene, cymene, d-terpineol, etc.

COLLECTING AND PREPARATION: The fresh twigs as a tincture, Method A. The whole branch can be hung and dried and chopped for tea. For internal use, other than bladder infections, roast the stems slowly in an oven or a frying pan to evaporate most of the volatile oils. This leaves the astringent functions intact.

MEDICINAL USES: The tincture of fresh twigs is second only to the similar preparation of Thuja in its antifungal effects. It is effective in all skin fungus infections, including ringworm, jock itch, athlete's foot, and the other tineas. Apply several times a day for several weeks; the same procedure is useful for pets as well.

A tea of the green twigs will inhibit urinary tract bacteria when they are the cause of urethritis and cystitis; combine with Manzanita, Yerba Mansa, Shepherd's Purse, or Hollyhock. See Formula #7 (pg. 138). The oils are excreted in the urine and retard bacterial growth; if used for extended periods, this tea may irritate the kidneys; in any case, it is not appropriate for those with any form of kidney disease.

The roasted herb is an old California Indian treatment for dysentery or diarrhea, being both astringent and mildly antispasmodic. Use it alone, or with Chaparro Amargosa if you suspect amoebic or giardiac causes, with Echinacea if bacteria or food poisoning is the cause. The Cahuillas and Pomos applied the roasted tea to injured, badly abraded, ulcerated, or burned skin to promote healing and granulation.

OTHER USES: The ground dried twigs make a lovely incense, and may be layered in stored woolens to repel insects.

DESERT LAVENDER

Hyptis emoryi   Labiatae
OTHER NAMES: Bee Sage, Yerba del
Becerro, "Salvia Real"

APPEARANCE: This is a large, attractive greyish shrub, compact in form and up to eight or nine feet tall. Desert Lavender is a plant that forms stands and colonies, either solid masses in gravelly flats or, along the edges of arroyos, all strung out like a silver fringe. The flowers form violet balls, strung like fluffy beads along their terminal stems, sometimes nearly covering the edges of the bush. The overall appearance is that of a tall, shrubby combination of Horehound, Garden Sage, and Catnip. If you aren't quite sure about the plant in front of you, crush the leaves—they smell of lavender and Pine.

After a desert rain, you can locate a stand by your nose. A little too low in the valley and all you smell is Chaparral, but if you walk up the center of an alluvial fan and out of the lower junk, you may get a whiff of a sweet-sachet smell. Follow it upwind until you come to the stand of Desert Lavender.

HABITAT: Frost-free washes, arroyos, and gentle hills throughout the Colorado and Mojave deserts as well as those in Arizona; south into Baja California and Sonora. Again, look for them in the warmwind foothills and bajadas above the Chaparral (*Larrea*).

CONSTITUENTS: Terpenoids, flavonoids, B-sitosterol, oleanolic acid.

COLLECTING: The flowering and leafy branches, dried, Method A.

STABILITY: Up to a year; the older it gets, the less pleasant the scent.

PREPARATION: Standard Infusion (2–4 oz. as needed); drink cold for hyperacidity, hot for everything else. The tincture, Method B, 1:5, 50% alcohol, ½ teaspoon in cold or hot water depending on use.

MEDICINAL USES: Like the true Sages, Desert Lavender is an anesthetic to the esophagus and stomach that, if drunk cold, also inhibits hypersecretions of red, inflamed, and irritated stomach linings. It is a first-class tea for hangovers with the usual nausea, and for that horrible period after a bout with gastroenteritis ("stomach flu") when your mouth tastes like a hog trough looks, and food sets in your stomach like a snake crossing the Ventura Freeway at five in the afternoon.

Desert Lavender has distinct hemostatic effects as well as astringency and was used by desert Indians for staunching heavy menstruation and bleeding hemorrhoids.

DESERT SENNA

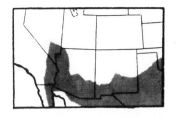

Cassia spp.   Leguminosae
OTHER NAMES: Partridge Pea, Sensitive
Pea, Té de Sena, Brico, Bush Senna, etc.

APPEARANCE: The most characteristic part of these plants also indicates that it is time to pick them: the golden-yellow flowers. These are rather showy and five-petaled, with at least one petal irregular; this gives Senna a rather violet-like appearance. The plants are obviously members of the Pea family, with their pinnate leaves, ranging from as few as two leaflets (*C. roemeria* of Texas) to many, always paired on the central leaf stem and often a bit smelly. Senna leaves never have a terminal (odd) leaflet.

Still, it's those flowers, usually in racemes, hinting at several shades of yellow and orange, rising up from the bright green upright stems with the pinnate leaves, that tell you it's probably Senna. The flowers are from one to one and one-half inches across, with a tendency to curl inward. Senna form flat pea pods, but it's best to gather them in flower, when the leaves are strongest and confusion with other plants unlikely.

HABITAT: *C. leptocarpa* is a big, handsome herbaceous Senna with sharply pointed two- to three-inch leaflets. It grows in broad alluvial valleys or along streams below 5,500 feet in the southern half of New Mexico and Arizona. Partridge Pea (*C. fasciculata*) is a common and pretty annual, blooming in July and best picked when two to three feet high. It grows from southeastern New Mexico through to Arkansas and eastwards. *C. wrightii* and *C. lepadenia* are somewhat similar annuals, found in the southern and lower elevations of Arizona, New Mexico, and Texas, southwards into Mexico. The most common Senna of our Sonora deserts is *C. armata*, which can be seen blooming in April and May of wet years, from the Mojave to those of Arizona and north to Las Vegas, Nevada. This Desert Senna has barbed leaf tips, somewhat irregularly paired leaflets, and thick, green-barked branches up to two feet high.

CONSTITUENTS: Various sennosides and free anthroquinones; our western Cassias have had little natural products chemistry research.

COLLECTING AND PREPARATION: Gather the leafing branches and dry, Method A; strip off the leaves when dry and discard the twigs and branches. Use the leaves for tea (a cold infusion tastes better and works as well as the hot), 2–3 grams in a cup of water is a safe trial dose. The Sennas may vary a lot in their strength.

MEDICINAL USES: First, last, always, a laxative. The tea is a useful cathartic, producing soft, watery evacuations. It does this by stimulating stomach and duodenal secretions, exciting peristalsis in the small intestine, and finally exerting its greatest action on the colon. It is most applicable for the person with dry mouth and skin, no inflammation, piles, ulcers, or hemorrhoids (or other complications), and with frequent marbly-mucousy constipation. Drunk in the middle evening, it will work by morning. If taken alone, however, it may give you nausea, cramps, and the sensation of a herd of gerbils running the Boston Marathon across your transverse colon. It is necessary to combine the 2–3 grams of leaves with some strong and pleasant antispasmodic herb, in order to get the laxativity without the negativity. If you are being a purist and only combining the botanicals I have somewhat arbitrarily placed in this book, combine the Senna with Chimaja, Desert Lavender, Marsh Fleabane, Pennyroyal, or Pineapple Weed. Otherwise, take the tea with ¼–½ cup of black coffee (that's right, coffee), or throw some dry Ginger, Cinnamon, Cloves, or strong Peppermint in with the Senna and brew together.

Senna will not cause rebound constipation, like Rhubarb or Yellow Dock can, but it definitely can contribute to a laxative dependence if constantly used. Laxatives, like sedatives, always have an idiosyncratic effect on different persons. If you find that one works nicely for you when you need it, and you need it frequently but don't wish to develop a dependence, mix it with others. Try one-half the effective dose of the leaves, add ¼ teaspoon of Yellow Dock and a scant teaspoon of Licorice Root from the health food store. This broadens the type of laxative effects to include not just the raw irritation of Senna but stimulates bile and secretory functions as well.

CONTRAINDICATIONS: Not for pregnancy (it can cause sympathetic uterine contractions), not for nursing (it passes into the milk), not for use with gastrointestinal inflammations (it will only aggravate them), not for use with ileus, volvulus, or any possibility of intestinal blockages.

DESERT WILLOW

Chilopsis linearis   Bignoniaceae
OTHER NAMES: Mimbre, Flor de Mimbres,
Catalpa linearis

APPEARANCE: When not in bloom, Desert Willow is an upright, willowy shrub and small tree, with many long, thin leaves that are as wide as a pencil, six or seven inches long, and slightly curved. The older bark is dark brown. The seed pods are similarly long and slender, splitting lengthwise when mature and persisting on into the dry summer when the leaves have already fallen. The flowers are large, showy, and trumpet shaped, up to one and one-half inches long, blooming in May and June, later if there are good summer rains. They form many terminal clusters of pink to lavender colors and have a heavy, musky fragrance.

HABITAT: Desert Willow is a water indicator in the desert and is classed as a phreatophyte; if you see some growing, there is water not too far below the surface during part of the year. It frequents desert washes, grasslands, and valleys from 1,500 to 5,500 feet. Its range is from Redlands east through the Mojave and Colorado deserts, the deserts of Arizona, and along the Pecos and Rio Grande basins in New Mexico and Texas. With overgrazing and poor soil management, it has been extending its range north and can be found occasionally in the Panhandle of Texas, Oklahoma, and even into Kansas.

CONSTITUENTS: The flowers contain anthocyanins, the seeds contain trienoic fatty acids, the leaves and branches contain alkanes, squalene, and at least one piperidine alkaloid; the bark and wood have varying amounts of tecomin.

COLLECTING: The flowers are gathered and dried carefully, Method B; they are high in nectar and spoil easily. The leaves and bark are sturdy, so dry in whatever fashion the particular plant dictates.

STABILITY: The flowers will last up to six months, the rest, forever.

PREPARATION: The leaves and bark tinctured, Method B, 1:5, 50% alcohol or powdered for topical or tea use. The flowers may be brewed for a simple tea or powdered for poultice.

MEDICINAL USES: In northern Mexico, where it is abundant and widely used, the flowers are made into a tea and a moist hot poultice. It is used for hectic coughing with a flushed face and the sensations of chest and lung tiredness with a rapid, thin pulse.

The powdered leaves and bark are an excellent first aid for the desert hiker, or at home for the desert city hiker, dusted liberally on scratches, rock scrapes, and the like. The tincture works similarly, of course, but stings.

Most important, perhaps, is its anti-fungus and anti-Candida use. As a tea (strong infusion, 2–3 oz. twice a day) or the tincture (¼ – ½ teaspoon twice a day), it inhibits Candida suprainfections. After antibiotic therapy or anti-inflammatory drugs, many people get episodes of foul burping; acid indigestion; loose, abnormal stools; hemorrhoids; or rectal aching; and even varicose veins. This can be the result of a subtle and intransigent *Candida albicans* infection in the upper and lower ends of the intestinal tract. Desert Willow, alone or combined with Echinacea and/or Chaparro Amargosa, can offer a sensible treatment.

OTHER USES: Someone out there with patience and an aesthetic perversity should try extracting the scent of the fresh blossoms. It is heavy, sweet, and sexy.

ECHINACEA

Echinacea angustifolia   Compositae
OTHER NAMES: *Brauneria angustifolia,*
Kansas Snakeroot, Black Sampson,
Purple Coneflower, Spider Flower

APPEARANCE: Echinacea is a long-living perennial, averaging one to one and

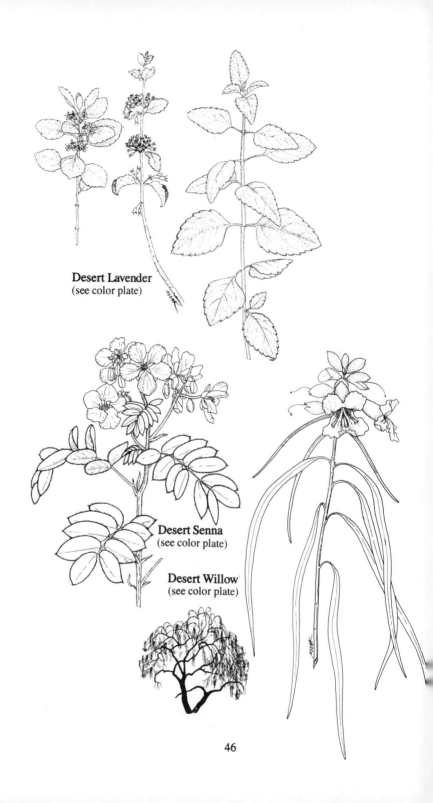

Desert Lavender
(see color plate)

Desert Senna
(see color plate)

Desert Willow
(see color plate)

46

one-half feet in height. In May through July it produces from one to ten coneflowers on thick stalks, light purple to light pink, petaled with a brown to dark brown, prickly sunflower-like center. The petals (ray flowers) are darker when emerging from the flower head and distally lighter, as if faded from the sun. The whole plant is prickly and hairy. The mostly basal leaves are narrow and up to four inches long. The stiff seed heads are visible into late fall, and after the petals fall off, they are the main characteristic of the plant. When spending hours driving and hiking in the late summer Texas Panhandle heat, staring earnestly up hillsides and road cuts for the elusive little black balls of *Echinacea angustifolia*, I dream all night . . . about driving and hiking in the late summer Texas Panhandle, etc.

HABITAT: This, the strongest species of Echinacea, is adapted to the limestone hills and arroyos of the western plains, from central Texas to the New Mexico border, up through the Panhandle, the white hills of Kansas, the Sand Hills of Nebraska, through eastern Colorado (scarce), Wyoming and Montana, and eastwards. The best medicine grows in these limestone soils of the western prairies and the lower drainage of the front range and the Canadian River.

CONSTITUENTS: The polysaccharides heterotri-and-heterotetraglycans, echinacein, echinolone, echinoside, echinacin B, sesquiterpenes and diterpenes, inulin, two isomeric 2-methyltetradecadienes, several conjugated forms of caffeic acid, etc.

COLLECTING AND PREPARATION: The fresh root tincture (gathered from August to November), Method A. The dry root and seedheads (gathered as early as June), tinctured 1:5, 70% alcohol, Method B. The seed heads or flowers, dried Method B, for making a salve, Method A.

Since this is a special plant, with active constituents that have differing solvents, here is an Echinacea-only method of making the most impeccable extract possible. Percolate as if for a 1:3 tincture, using 80% alcohol, and macerate in the percolator for 48 hours, instead of the usual 12–24 hours. Draw the 1:3 strong tincture and put it aside; this is Extract #1. Remove the marc, add 5 parts of hot water to it in the top of a double boiler, steep over boiling water for two hours, remove from heat, cool down, and squeeze the fluid from the marc, which is then discarded; this is Extract #2. Evaporate this second extract over the boiling water until 2 parts in volume. Combine both extracts to form a 1:5 tincture in which both the aromatics and the mucopolysaccharides are in maximum solvency.

Hypothetical Example. You have 5 ounces of dried root and seedheads; you break it apart, grind it down, press the powder into a measuring cup, and find that it takes up 10 ounces of volume. It will therefore hold 10 ounces of menstruum in the percolator and, as you need to draw 15 ounces of finished strong tincture (1:3) and will lose 10 ounces to the herb in the percolator, you need to make 25 ounces of menstruum. At 80%, that means mixing 20 ounces of pure alcohol with 5 ounces of water. You moisten the powder, wait an hour, pack it carefully in the percolator, set it (covered) aside for two whole days, pour the remainder over the column, and draw 15 ounces of over-strength tincture. Then

47

you remove the marc from the cone (which is passively holding 10 oz. menstruum), add the 25 ounces of hot water (5 parts), and mix them in the top of the double boiler. Boil the water in the lower section, steep the muddy grey gruel for two hours over the steam bath, remove, cool, and squeeze through a cloth. You throw away the tired old herb. You *now* have in front of you one bottle holding 15 ounces of evil-looking reddish brown Echinacea tincture (1:3), and another volume of approximately 2½ cups of grey, milky, slightly alcoholic soup. Put the soup pack over the double boiler and evaporate it over the steam for several hours until the 2½ cups is reduced to 10 ounces in volume (2 parts). Combine the 10 ounces and the 15 ounces, and you now have a mucopolysaccharide-amplified tincture (1:5) of *Echinacea angustifolia*. The five ounces of dried herb is now digested into 25 ounces of *really* wicked-looking tincture. This all may seem complicated (and it is), but the good Echinacea you gathered is so damned useful that it warrants this labor and respect.

STABILITY: The dry, washed, uncut root and flower head will last for 12-18 months with reasonable strength. Commercial cut-and-sifted or powder has a short shelf life, so if you can't gather it, try to obtain the whole root from someone else.

MEDICINAL USES:

Immunologic: Echinacea angustifolia, internally as well as topically, stimulates the body's defenses across a broad spectrum. Firstly, it speeds the responses of white blood cells during their attacking and digesting of bacteria, toxic immune-complex proteins, and the larger viruses that cause influenza, herpes, and vesicular stomatitis. With Echinacea present, their ability to roughen up bacterial membranes for proper digestion is increased (opsonization), as is the properdin pathway of complement. Complement proteins are produced by defense cells as an anchor for antibodies . . . they stick to the bacterial membrane at certain intervals so the antibodies can punch a hole through the membrane and "neutralize" the microbe. In addition, Echinacea seems to speed up the rate of lymphocyte replication. Once the body recognizes a previously encountered bacteria, virus, or mold, lymphocytes must be cloned that manufacture the specific antibody for that particular organism; this happens faster when Echinacea is present in the bloodstream.

Further, with a body of toxic "battle" trash floating around in the infected or injured tissue, made up of dead microbes, dead white blood cells, and pieces of dead cells, the scavenging leucocytes that have to clean out this pus seem to be able to clean up better. This improves overall defenses, and resolution occurs more quickly.

Resolution is critically important in the healing process. Most tissues are made up of two basic types of cells, mesenchymal and parenchymal. Mesenchymal cells form structure, glue, and vessels; they are slow metabolically, durable, and repair themselves perfectly well in a polluted fluid medium. Parenchymal cells perform function by secreting, communicating, contracting, digesting, and have a higher rate of metabolism and spend life energy performing these functions and are more delicate and highstrung. They also replicate and regenerate very poorly in a polluted environment. The sooner fluids are cleaned out and the various cells begin healing, the better the ratio of structural and functional cells is maintained. To sum up my case, the quicker the junk is removed, the less scarring occurs and the better the flesh works after healing.

Tissue Repair: Echinacea, as explained, speeds up the repair of tissue damage in general. Cells live in a compacted, jello-like fluid, a starch gelatin that pulls water out of cells as it is produced by combustion; glucose and oxygen, when burned, release water and carbon dioxide as waste products. This dehydration allows orderly streaming of liquids between entering and exiting blood capillaries, with the solid trash drawn electrically and passively away to the lymph capillaries. This intercellular "jello" is called a mucopolysaccharide hydrogel. If too much protein leaks out of the capillaries from extended inflammation, or if too many cells burst and die from an injury, or those bacteria that secrete enzymes that can dissolve the gel move into the tissue, then cells start to expand, and their waste products float around them. The overall congestion then blocks orderly lymph drainage, and the garbage goes uncollected. This is the main cause of the edema and swelling that come with injuries and infections.

Echinacea, particularly the polysaccharides therein, helps to hold the starches together longer, protects the hydrogel against dissolving, and aids speedy regeneration of the tissues. Applied topically in a salve or tincture form together with internal ingestion of the tincture (¼ – ½ teaspoon every several hours), Echinacea helps limit the swelling and edema of a variety of problems, from hemorrhoids, contusions, and sprains to the stings of bees, wasps, conenoses, and mosquitoes.

Connective Tissue Swelling: When cartilage is injured, the dense gristle must be softened in order to facilitate increased blood supply, white blood cell housecleaning, and general repair. Healing tendons, ligaments, and muscle sheaths are like a snake that has just molted: soft, delicate, and easily injured. Re-injury during this time can become chronic tendonitis or bursitis, with the cartilage cells in a constant state of congestion and inflammation, chronically secreting the enzymes that soften up their environment and constantly painful. Use up to ½ ounce of the tincture a day to support the density and tensile strength, until the swelling reduces and the pain leaves. This is great for tennis elbow, skier's knee, and jogger's ankle. Although not as strong as prescription topical medication, the regular use of small amounts of the tincture internally (30 drops, five times a day) helps decrease corneal opacities and some of the interocular pressure of glaucoma.

Internally, Echinacea combines well with Red Root (see *MPOMW*), Chaparro Amargosa, Desert Willow, Cypress, and Hollyhock for infections; Elephant Tree, Yerba Mansa, and Milk Thistle for tissue repair; and Milk Thistle and Red Root for connective tissue swelling.

OTHER USES: The essential oils of the seeds and to some degree the root inhibit the maturing of grain beetles and mealworms.

ELEPHANT TREE

Bursera microphylla   Burseraceae
OTHER NAMES: Torote, Torote colorado

APPEARANCE: A striking, small tree or shrub, from five to ten feet high in the U.S. but twenty feet or more in height in western Mexico. The name comes from the thick, enlarged trunk and main branches. The translucent, paper-peeling bark is a butterscotch-yellow. The smaller branches and twigs are not thickened and have a copper or reddish-brown hue. The dark green leaves are pinnate and Mesquite-like, persisting throughout the year except during extreme drought or after a pronounced freeze. The fruit is small, succulent, and light purple or pink, each containing a single yellow seed; they persist through the summer and winter, interspersed here and there with the leaves. The whole tree has a strong tangerine-incense fragrance. Even where abundant, such as central Baja California and western Sonora, it is a strange and anomalous plant, quirky but friendly. It reminds me of a fat and jaundiced boa constrictor, dressed up in a poorly fitting tree suit.

HABITAT: Southwestern Arizona and the eastern, dry hillsides of the Anza-Borrego State Park in California, south in greater abundance around the Sea of Cortez in Baja California and Sonora. In the U.S. it is found in small stands in the rocks of dry, low mountains and their alluvial fans, basking in warm air currents and out of the frost. In Arizona it is found as far north as the Casa Grande Mountains, as far east as near the Kitt Peak Observatory, and west to the Telegraph Mountains near Yuma. If you simply wish to observe the plant, there is a marked (and protected) stand south of Ocotillo Wells in the Anza-Borrego, and similarly marked (and protected) stands in Organ Pipe National Monument, Arizona.

The rest of our stands are found in the nooks and crannies of some of the most remote desert mountains imaginable, such as the Tinajas Altas, Sierra Estrellas, Growler, and Mohawk ranges. The only reason I am writing about this rather rare (in the U.S.) tree is that you don't need to take any more than a small branch, some leaves, and the exuded gum . . . nothing life-threatening or substantial for the plant. Besides, outside the California stand and a large colony east of Yuma, you have to dodge heat-seeking missiles (Luke Air Force Bombing Range), game wardens and terrible roads (Cabeza Prieta Game Range), Smokey the Bear (Organ Pipe), and cautious Elders (Papago reservation) . . . and some ranchers. Although actually rather abundant in Yuma and Pinal counties, where Elephant Tree grows is HARD to get to, and then you have to walk up a mile through a hot, rocky bajada and then scramble up boulders. Those of you willing to go to these lengths to gather some of their foliage and gum are not likely to offer them any danger. To the rest of you that find this sounds like a B-movie Vision Quest, just wait until the next time you take a vacation to Rocky Point, Kino, or Quaymas and gather some where it is an abundant plant. As for me, give me the east slope of the Growlers, a full moon coming up, the Elephant Trees to my back and missile practice down in the valley.

CONSTITUENTS: Burseran, B-sitosterol, deoxypodophyllotoxin, myrrins, and several lignins with experimental anti-tumor activity.

COLLECTING AND PREPARATION: Bark and twigs, chopped for a fresh tincture, Method A; leaves dried for tea; the resinous gum, tinctured, Method B (macerate), 1:5, 75% alcohol, or for burning as incense on charcoals.

MEDICINAL USES: The tree, a similar biotype to its relative, myrrh, and with similar constituents, has similar immunologic stimulation. It will increase phagocytosis, both the numbers and quality of serum white blood cells (PMNs), as well as the granular streaming, when viewed under darkfield live blood analysis. There is nothing magical here, but it means if you are tired, rundown, and getting a little sick a lot of the time, the tincture of the bark, gum, or the leaf tea helps strengthen your resistance while you are under stress, especially if you couple the Elephant Tree with a little Echinacea, Red Root, Cypress, or Hollyhock.

The aromatic oleoresins are primarily excreted in the urine and mucus as intact waste products; as such they inhibit bacteria and other microbes, stimulate the scavenging of white blood cells in those tissues, and increase the softening and expectoration of bronchial mucus. Elephant Tree would be classed, therefore, as an excretory disinfectant, mucolytic, and immunostimulant. As it, like myrrh, is strongly astringent as well, the various preparations are very useful in treating gum and mouth inflammations.

The doses for internal use should be 20–30 drops to five times a day for the tinctures, a mild tea of the leaves brewed, from boiling water, long enough to be warm and slightly bitter, up to four times a day.

OTHER USES: The gum for incense, like copal.

CONTRAINDICATIONS: Kidney disease (may induce inflammation), pregnancy (may overstimulate uterine blood supply), and necessary immunosuppressant therapy.

ENCELIA

Encelia farinosa   Compositae
OTHER NAMES: Incienso, Brittlebush

APPEARANCE: Most of the year this plant is a low, rounded mass of grey-white leaves, a foot or two high and several feet in diameter. In the spring (February to April) it sends up tall, skinny stalks that flower into bright yellow daisy blossoms, hovering like butterflies over the silver leaves. This is one of our more spectacular wildflower displays. For years, traveling east or west along I-40 or I-8 through the lower deserts, I would see these plants and think that they were White Sage (*Salvia apiana*), a favorite plant of mine. Year after year, even though I knew better, I would stop, crush a leaf, get no smell, and mutter grumpily. In the spring, when blooming their bright yellow excess, I was never fooled and kept on driving while muttering grumpily.

HABITAT: Everywhere along the roads and valleys of the lower deserts of California and Arizona, making its way into Nevada and Utah, down into the western states of Mexico; from sea level to 3,000 feet. You can't miss it.

CONSTITUENTS: The leaves and roots contain various unidentified diterpenes and sesquiterpenes; the gum contains benzoic, cinnamic, and abietic acids. Never completely analyzed, to my knowledge.

COLLECTING: The heavily leaved branches, broken off near the root, dried, Method A; strongest in late fall and winter. The gum, usually hardened by the late summer, collected in any handy container. Always check around the base of the plant, as the gum often drips down onto gravel or leaf detritus and hardens into little yellow tears easily seen. The soft, runny pitch is viscose and sticky and can only be removed from utensils, flesh, and clothing by sand blasting or a discreet, "clean," thermonuclear device.

PREPARATION: Leaves and stems a Strong Decoction, two to three ounces. The gum dissolved in some warm lard, made into a salve, Method B, or simply chewed up.

MEDICINAL USES: The tea has a strongly bitter and slightly numbing effect. It is a well-known folk remedy in northern Mexico for arthritis that is aggravated by cold and damp weather, 2–3 oz. of the decoction three times a day while the pain is acute. Cahuilla and Mojave Indians used it for a mouthwash and retained gargle for tooth and gum pain; it works quite nicely. Also, a few sips of the hot tea will efficiently induce sweating when you are trying to break a cold or flu fever in order to get on with the business of getting well . . . similar to Boneset (*Eupatorium perfoliatum*) or Yarrow. The gum, dissolved in lard or made into a salve, can be applied to the chest, neck, or throat to stimulate expectoration and loosen up thick, intractable mucus that lingers from bronchorrhea or bronchitis, especially if moistened, fairly hot towels are put over the ointment. A single lump or two of the gum works similarly.

OTHER USES: The gum is widely used in northern Mexico as a church incense or smelly-cigar incense, dropped on coals like Copal or frankincense. Delightfully elegant.

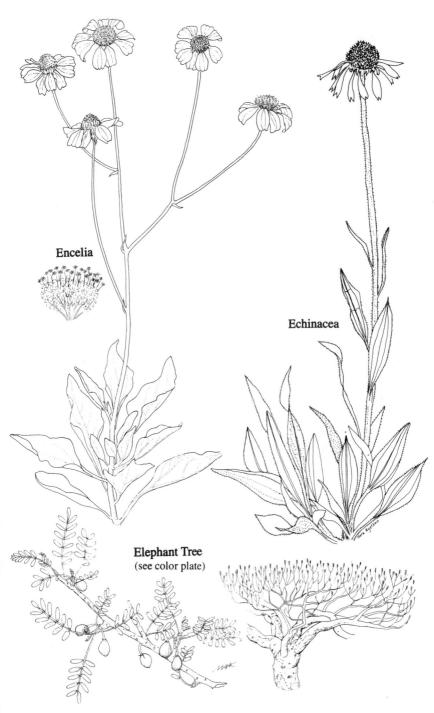

Encelia

Echinacea

Elephant Tree
(see color plate)

EPAZOTE

Chenopodium ambrosioides   Chenopodiaceae

OTHER NAMES: Pazote, Epazote de Comer, Mexican Tea, American Wormseed

APPEARANCE: Epazote is a scruffy, nondescript weed, usually an annual, sometimes a short-lived perennial. The leaves are irregularly toothed and light green but frequently having reddish or purplish splotches. It closely resembles its relative, Lamb Quarters (*Chenopodium album*) but without the latter's lime-green, frosted look. The whole plant, mature by early summer, is from two to four feet high, either sparsely or heavily leaved or branched. If growing in side-walk cracks in East Los Angeles, Barstow, or Phoenix, however, it may be a splayed out, much trodden-on orphan. The flowers mature into seeds, but without any appreciable change in appearance, sometimes turning slightly red, sometimes not. They crowd around the upper stems in dense clusters like Amaranth. The stems are variously ridged, stiff, and woody. The whole plant has a typical pungent-gross scent, kind of stinky, kind of nice.

HABITAT: Unpredictable. A native of Mexico and Central America, Epazote can be found most frequently in vacant lots of Sunbelt cities and salt marshes and dirty suburban drainage ditches. It is especially abundant in southern California and along Colorado River waterways but can otherwise be found anywhere.

CONSTITUENTS: Ascaridol, p-cymene, 1-limonene, d-camphor.

COLLECTING AND PREPARATION: The whole plant in full seed, with as many leaves as possible. Strip the seeds off to dry separately; otherwise, they fall all over the place. Hang the remainder, Method A. Strain the seeds from their chaff when dry, strip the dried leaves off, and discard the hard, brittle stems. The leaves are used for tea or spice, the seeds are best taken in capsules after grinding enough for immediate use. The seeds may also be used in cooking, but they are more pungent than the leaves; so use sparingly.

MEDICINAL USES: Epazote has long been known as American Wormseed in the drug trade; although the official designation is *C. ambrosioides,* var. *anthelminticum,* the various varieties (if varieties they be) are virtually identical and inter-breed undeterred by botanists or pharmacists. Except during pregnancy, the seeds are a safe and reliable vermifuge for roundworms, *Ascaris lumbricoides,* a usually benign, human-specific parasite that, nonetheless, may sometimes get out of hand. It is estimated that one out of four human beings have them, usually without disease symptoms. If they are present and you have a firm diagnosis from a physician, NOT by muscle testing, biokinesiology, or iridiagnosis (reasonable tests in some circumstances but notoriously erroneous in overdiagnosing *real* squiggly-squirmy parasites), then here is the proper way to use Epazote seeds for their removal. Take a good strong laxative in the evening (Cascara Sagrada, Senna, or Yellow Dock), fast for the first twelve hours the next day, take two grams (four #00 capsules) of ground seeds with a glass of Mesquite or Gum Arabic mucilage or a cold infusion of Hollyhock or Marshmallow, then wait two hours. Follow that up with a full dose (groan) of castor oil, then watch those little varmints come out.

Although Epazote is sometimes recommended for hookworms, I would not advise such an approach. If you don't kill them outright, they can cause increased damage to the intestinal lining. Go for a regular pharmaceutical; present medications work fine and are more efficient at removing these debilitating parasites.

Frequently touted as an abortifacient or menstrual stimulant, Epazote is potentially toxic in large doses. There have been many cases of poisoning from taking large amounts of the tea or the distilled oil, and it seldom works. On the other hand, all parts of the plant are effective externally for fungal infections, barber's itch, athlete's foot and ringworm, as well as being somewhat antibacterial.

OTHER USES: The leaves and the seeds are a classic Mexican bean spice. It is called for in many traditional recipes, both to reduce the flatus levels and to jazz up the taste.

ESCOBA de la VIBORA

Gutierrezia spp.   Compositae

OTHER NAMES: *Xanthocephalum* spp., Matchweed, Snakebroom, Broomweed, Yerba de la Vibora, Collálle

APPEARANCE: The several species, *G. sarothrae, G. californica,* and *G. lucida,* are all similar in growth as to be virtually interchangeable . . . at least to us non-botanists. Escoba (for short) is a small, many-stemmed, upright shrub with numerous narrow leaves and many little yellow flower clusters, three to eight blooms in a bunch, with small, scaly, overlapping bracts. The plant has a resinous or waxy finish, and when crushed it gives off a slight piney scent. Perennial, new growth and previous dead stems are intermixed in the same bush. An insignificant bush, reaching an insignificant height, with insignificant leaves and flowers, its appearance is so common and uninteresting as to approach . . . insignificance.

HABITAT: High and dry slopes, mesas, bajadas, river bottoms . . . you name it, there is probably some Escoba there. It is found in abundance in all the states and deserts and suburbs of our area, from 1,500 feet to 9,000 feet. It is the recipient of the largesse of overgrazing, above the Chaparral (*Larrea*), mingled with and above the Mesquites, into and above the Juniper/Piñon belt, into the Sagebrush. I even saw some growing and flourishing alongside some *Arnica cordifolia* at 9,200 feet in southwestern Colorado.

CONSTITUENTS: The main active constituent is a diepoxyde, although it contains at least twenty different flavonol methyl esters. A group of plants to drive the chemotaxonomist crazy.

COLLECTING: Clip the flowering stems together into little bundles, band; dry, Method A or in a paper bag.

PREPARATION: Steep in a bundle in a quart of boiling water for one-half hour, cool, remove bundle (discard it), and add to a hot bath.

MEDICINAL USES: The hot bath is an efficient and pleasant treatment for arthritis and rheumatism, sore muscles, and hyperextensions. Like other palliative treatments, it won't cure anything, but it is effective in reducing inflammation and pain. The bath can be repeated as often as desired. Be warned, however, if you have arthritis; it works so well that you may need to gather dozens and dozens of little bundles to last for the fall and winter months. Not that it isn't abundant; it's so abundant that you may not take it seriously. A respected, almost revered *remedio* among Hispanic New Mexican and Arizona folks, where a tea of the herb is usually drunk while bathing in it. The infused tea can also help the nausea from hangovers or a simple stomachache and will slow heavy menses.

Although by no means a heroic medicine, it is common, safe, and may sometimes work so well for joint inflammations as to supplant salicylate (aspirin) treatment or be used alternately with other pharmaceutical drugs to decrease their side effects.

EUCALYPTUS

Eucalyptus globulus and others   Myrtaceae

OTHER NAMES: Blue Gum Tree

APPEARANCE: Eucalyptus is Eucalyptus. A tall, rapidly growing tree, *E. globulus* makes up 95 percent of our Eucalyptii, characterized by the brown-grey, peeling, deciduous bark, the white flowers maturing into the typical warty little seed-muffins that litter the grounds of so many of our Sunbelt parks, and the . . . scent. The mature leaves are long, sickle-shaped, and hang in abundance from the stems of the trees; younger leaves, pubescent silvery-blue, are more rounded and pop up from all over the immediate area of the trees like demented Spearmint.

HABITAT: Wherever it doesn't freeze much, and someone has planted them. Although I have seen many wild-growing Eucalyptus trees, and old established groves of twenty trees planted in 1930 have mysteriously become 200 trees, almost no botanists list the plants as naturalized. Although only a botanist by function (not by title), I hereby declare *Eucalyptus globulus* a Naturalized Tree.

CONSTITUENTS: Eucalyptol (cineole), various volatile aldehydes, phenols, phellandrene, piperitones, cerylic alcohol, resinoids, eucalyptic acid, tannins.

COLLECTING AND PREPARATION: The best leaves for our use are the younger, less sickle-shaped or overtly round opposite leaves of the offshoots and young branches. Older leaves or those of individual trees growing in marginal areas have less of the Eucalyptus scent and contain more phellandrene and piperitone constituents; these are moderately toxic and sometimes cause headaches. A simple hot infusion of several leaves is the most practical way to drink the herb, with a stronger infusion being used for external washing.

MEDICINAL USES: Eucalyptus tea made from good, aromatic young leaves is sure to induce sweating, stimulate kidney function, and inhibit microorganisms in the bronchial and sinus mucosa. It stimulates mucus secretions in all the

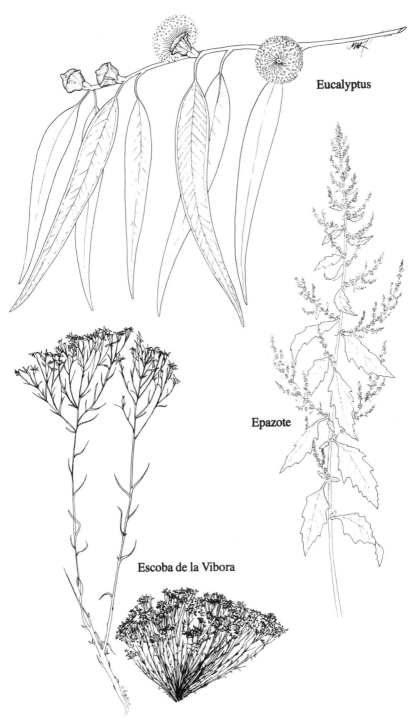

Eucalyptus

Epazote

Escoba de la Vibora

57

mucosa; the aromatic gases are exhaled from the bloodstream into the lungs and out the respiratory tract, actively remoistening the membranes and inhibiting the bugs growing on them. For those suffering from stomach fermentation, where every burp is sulphurous and foul and food tastes dreadful, three or four cups of tea during the day will almost always kill off the yeast or bacteria in the upper stomach lining causing the distress. Your breath may end up smelling like Vicks Vaporub, but it will no longer resemble demon flatus. Topically, the strong infusion will kill most bugs on the skin, both infectious and benign, and it is a good cleansing wash, both astringent and antimicrobial, before bandaging or dressing a minor or moderate wound. Many books have recommended little necklaces of the pods for hanging around the necks of flea-bitten pets. It has been my experience that, if the pet can't gouge off the necklace (they don't like the smell), then all the fleas migrate to the poor animal's nether regions (where they would just as soon be, anyway).

HOLLYHOCK

Althea rosea   Malvaceae

APPEARANCE: This is the familiar garden flower, growing up to six or seven feet in height, the plant covered in those various white, pink, blue, and red flowers. In the Southwest, it grows as a biennial, farther north as a short-lived perennial.

HABITAT: This is an increasingly common urban weed, growing along curbs, vacant lots, and alleys from Albuquerque to Salt Lake City, Bakersfield to San Diego. It is most common in the older sections of our cities, towns, and farming areas. People planted it there at one time; now it has taken over.

CONSTITUENTS: Mucopolysaccharides (mostly xylan and glucosans), asparagine, fatty acid esters, dioxybenzoic acid, and an estrone-like compound.

COLLECTING: The roots in the fall or winter, peeled while fresh and any dark, woody sections cut out. The lighter the color, the better the medicine; tannish roots work well, just not as well. Cut and split into one-quarter- to one-half-inch sections, dry, Method B. Collect the leaves in early summer when the flowers are just starting to bloom, bundle by the long stems, dry, Method A.

PREPARATION: Both the leaves and roots are best prepared as a Standard Cold Infusion; this dissolves the active stuff and does not cook up the starches. The whole fresh root may be tinctured, Method A, if you wish a concentrated preparation. The tea may be consumed ad libitum, the tincture, the same.

MEDICINAL USES: The root or leaf tea is an old honored treatment for sore throats, stomach or duodenal ulcers, intestinal tract infections with vomiting and diarrhea, kidney and urinary tract infections, and as a douche for vaginal inflammations.

If you wonder why you've never heard of Hollyhock as a medicine, that's because its close relative Marshmallow (*Althea officinalis*) is the recognized species. Both plants have the same constituents and are often harvested together, but Hollyhock grows here. Since it is a common city weed and I have only seen one Marshmallow plant growing wild in all the western states, and since few people are aware of its usefulness, Hollyhock is the species included in this book.

The powdered roots or leaves make a first-rate poultice for obstinate skin infections. Grind some up in a blender or hand mill, rub it through a fine sieve, and save the too-large pieces for tea. Mix the strained herb with water, smear on some moistened gauze, place over the injury, and apply hot, wet towels (as hot as you can stand) for up to an hour, replacing the towels with hotter ones as they cool off. Repeat every four hours until the infection starts to localize and is on the way out of the tissues. This is all very nice and helpful too, but whole wheat flour (or even powdered, dried bananas) make a decent drawing poultice. The difference, and an important one, is that the mucopolysaccharides (the gooey mucilage) have been shown to possess an ability to increase the speed and ferocity with which white blood cells can grab bacteria, roughen up their surfaces, and eat them. To improve this action, try moistening the powder, not with water, but a tincture of Echinacea, Chaparral, the juice of Prickly Pear, or Indian Root.

The cold infusion, alone or combined with ½ part Echinacea root powder, helps to limit the damage that occurs from prolonged inflammation—that ulcerative, boggy degeneration far more harmful in many cases than the originating infection or irritation.

Stomach ulcers are especially benefited by Hollyhock, alone or with Echinacea, and the cold infusion should be used in recuperation from any abdominal surgery.

INDIAN ROOT

Aristolochia watsonii   Aristolochiaceae
OTHER NAMES: *A. brevipes, A. lassa,* Raiz del Indio, Indio, Arizona Snakeroot

APPEARANCE: This is an exotic little remnant in our deserts of an otherwise tropical family, a family so old that it evolved its insect-pollination schemes before insects could fly! A family that, nonetheless, had the genetic ability to evolve this, its only desert-dwelling species. It is a trailing, flattened semivine, with alternate arrowhead leaves, usually from one-half to one inch long. The flowers are purple-mauve little tubes that open into a spotted lower lip and mature into ballooned capsules. The whole flat little mat usually has a pronounced purple-brown hue, bland and almost invisible from a distance, elegant and distinct on close examination. Hard to find, but actually rather widespread.

HABITAT: From 2,000 to nearly 5,000 feet, from western Arizona to West Texas. Look for Mesquite, Ocotillo, Acacia, and Paloverde, walk up a sandy arroyo toward low foothills covered in masses of Agave, Prickly Pear, or even Saguaro, crouch down, and start checking the shrubby islands in the main wash or the side washes as they drain in, check under the Mesquites, check over there (see!) by the Escoba de la Vibora stand . . . If your back starts to hurt too much, your eyes get weary, and your arms are too scratched up to stand any more . . . and still no Indian Root, then go back to your nearest herbarium, memorize the way the specimen looks, get a specific locality, and start over again. You might find your footprint on one from the previous day, made while you were looking for the plant . . . it's happened to me.

CONSTITUENTS: Aristolochic acid (aristolochine), aristolactone, and at least one unidentified eudesmane alkaloid.

COLLECTING AND PREPARATIONS: The whole plant, including the deep tuberous root, fresh tincture, Method A, 1:5 (not 1:2); the whole plant, the root thinly sliced, dried Method B, and powdered for capsules or tinctured 1:10, 50% alcohol. If the stand is small and you wish to perpetuate the few plants, then the foliage above ground, although not as strong, has the same pharmacology and may be tinctured fresh, 1:2, or dried, 1:5. Capsule dosage is (whole plant) three to five #00 capsules a day; (foliage only) five to eight #00 capsules a day. Otherwise, the tinctures (however made), fifteen to twenty drops up to four times a day, or in combination with other plants. The tea is too disgusting to drink, so forget it.

STABILITY: The dried plant is fairly stable up to a year, the tuber, longer.

MEDICINAL USES: The main effects of Indian Root are 1) stimulation of salivation and stomach secretions for better digestion (tincture only); 2) stimulation of better metabolism of amino acids and proteins by the liver; 3) stimulation of better phagocytosis by white blood cells; 4) stimulation of sweating to break a fever state; 5) increase of arterial blood supply to the viscera and decreasing peripheral vasoconstriction and hypertension; and 6) increased speed of recuperation after a lengthy illness by aiding the liver to restore protein and transamination deficit.

This is a lengthy list of uses for such a little plant but they are the same uses formerly given to its close relative, Virginia Snakeroot (*Aristolochia serpentaria*), an even smaller plant. With fewer aromatics than the latter, Indian Root is somewhat less useful as a vasodilator and respiratory stimulant but is at least its equal for the uses mentioned above. Also, with less stimulating aromatics, it is less irritating to the intestinal tract and may be used with greater safety and for longer periods.

The digestive functions are simple enough. The tincture (or if you must, a little tea) is intensely bitter and, if taken a little before meals, aggravates and stimulates by reflex both salivation and gastric secretions. For older folks, heavy drinkers, and adrenaline-stress types, with decreased gastric function and a chronic dry mouth, maybe morning bad breath, hay fever or other inherited allergies, people who can't eat too much at once, don't like breakfast, are chronically constipated, have dry skin (you know who you are), a little Indian Root a few minutes before filling the top of the tract with food will result in a more orderly and better segmented group of upper-digestive reflexes, better food absorption, and sometimes a better lower digestive function as well.

As for the rest, clinical tests have shown that aristolochic acid (and the plants, like Indian Root, that contain it) speeds up the rate of granulocyte phagocytosis. What that means is, if you take a bunch of white blood cells, put them in a fake-blood broth in a petri dish, give them a terrific headache with chloramphenicol and antabuse so they get too sick to go to work (work being the killing and eating of certain bacteria), then they recover faster, eat the bacteria, etc., when you add some of the aristolochic acid to their media. The tests also showed that the white blood cells recognize certain types of tumor cells and kill them faster, too. As is often the case in cancer research, the people (not the petri dish) who took enough of the substance to be helped with their cancer got sick ten-ways-from-Sunday by the high doses of the aristolochic acid.

The mechanism is far more useful when given in small doses to those who need some functional stimulation, not to those with serious, stubborn, organic disease, much the way the whole *Aristolochia* has always been used, under the name of Birthwort, Virginia Snakeroot, Serpentaria, or Raiz del Indio. The whole plant stimulates the rate of oxidation by certain "hot" cells in the body needed for infection resistance. This includes white blood cells in general, but even more importantly, the kupffer's cells and hepatocytes in the liver, and specialized cells in the muscles and kidneys. The increased metabolization of these cells, their more efficient elaboration of defensive oxygen radicals, the better utilization of labile proteins, phospholipids, and amino acids and their return to the cells after recuperation (nitrogen deficit) make Indian Root, in small, frequent doses, a perfect adjunct in treating almost any self-limiting infection. A good regimen for doing this, for aiding the body's innate resistance (as opposed to killing the microbe directly), is Indian Root and Echinacea tincture with Hollyhock (or Marshmallow) infusion for the first half of the infection. See Formula #8 (page 138). Add a lot of Alfalfa tea to it for the recuperating period, along with some high quality dietary proteins for replacing the electrolytes *and* nitrogens depleted during the first half. You get well quicker and with less stress.

OTHER USES: This is supposed to be an antidote for snakebite. I, for one, have not and do not plan to test it out. If I *do* get bitten by a venomous snake (actually hard to do), I will walk (not run) to the nearest paramedic, thank you.

CONTRAINDICATIONS: Not for use in pregnancy or with major drug therapy (Indian Root's liver stimulation may alter the metabolism of the drug), although fine with antibiotics and the like.

JOJOBA

Simmondia chinensis   Buxaceae
OTHER NAMES: *S. californica,* Goat Nut,
Coffeeberry

APPEARANCE: This is a bush of low to medium height, complex branched and broadly dense, with thick and leathery round leaves bluish green in color. These are paired and form a distinctive glandular swelling around the stems. The plants are different sexes; females, bearing hundreds of acorn-size fruit, are usually in the more sheltered and moister sides of the colonies. The leaves persist year-round, and the whole plant resembles Manzanita or Silk Tassel, except for the opposite leaves, blue green foliage, and overall environment. A handsome, distinctive, and frequently dominant plant where it grows.

HABITAT: In California, from San Diego through Riverside and Imperial counties, all along the foothills of the Imperial Valley and Anza-Borrego; from Kofa Mountains and eastwards in most of southern Arizona. Jojoba is cultivated extensively for its oil, mostly in Yuma County. You will find it in gravelly, well-drained warm slopes from 1,500 to nearly 5,000 feet; it does not grow in cold high-mountain drainages. Those of you living in San Diego, Phoenix, or Tucson have it made . . . just drive out into the foothills.

CONSTITUENTS: Seeds and seed oil: C_{20} and C_{22} straight-chain monoethylene acids and alcohols as esters (forming a liquid wax); leaves and twigs: isorhamnetin, 3-rutinoside, isorhamnoside, and esters of benzyl alcohol and 2-phenylethanol-1; an unnamed alkaloid.

COLLECTING AND PREPARATION: The seeds (female plants) gathered in the summer, fall, or winter, either from the plant directly or from the ground below it. The husk can be removed by pulling it off with the edge of a pocket knife, although rattling these plants around a bit after they have dried awhile causes the husk to come off more easily. Most ground-harvested seed has no husk (this has the advantage of allowing squirrels to gather them first and probably propagate the plant through the seeds they gather and forget about); it does, however, increase your yield of poor seeds. For "coffee," roast them at 250 degrees until darkened . . . most of the day; for unroasted tea, crack them with a nutcracker and grind them down a bit in a blender or mill. The leaves are dried on the branches, Method A, and brewed as a Standard Infusion, cooled down a bit, two to four ounces as needed.

MEDICINAL USES: The leaves are a good tea for chronic mucous-membrane inflammation, ranging from chronic colitis, vaginitis, and hemorrhoids to stomach and esophageal ulcers. In Mexico it has been widely used as a folk remedy for asthma and emphysema, but it is more a matter of aiding the injured pulmonary membranes than addressing any underlying causes. A tea of the seeds will decrease inflammation in pharyngitis, tonsillitis, and various types of sore throat. Two to three ounces of the infusion drunk every several hours decrease

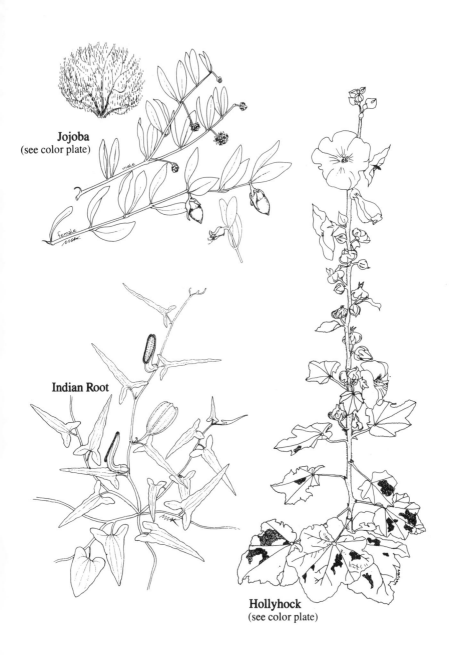

Jojoba
(see color plate)

Indian Root

Hollyhock
(see color plate)

the irritability of the bladder and urethra membranes in painful urination.

The seeds make a reasonable coffee substitute, not as good as the original but better than many pseudocoffees, except chicory, perhaps. Jojoba coffee has enough distinctive bite and character to be enjoyed on its own. The oil (actually a liquid wax) extracted from the seeds is the basis of a growing industry and explains the increasing cultivation of the plant. It is available from health food stores and co-ops; it ain't cheap, but it is an excellent scalp treatment for either dry, flaky dandruff or brittle hair. It is, however, being sold for everything imaginable and is allegedly put in all kinds of cosmetics. It is hustled as a treatment for all types of marginal distress, from toenail glitch to Advanced Irish Potato Famine.

CONTRAINDICATIONS: Some side effects may occur from the leaf constituents if taken by antabuse users.

JUNIPER

Juniperus monosperma and others   Cypressaceae

OTHER NAMES: "Cedar," Cedron, Sabina

APPEARANCE: There are two general types of Junipers: small trees with dark olive green scaly, leafless twigs and the high altitude spreading shrub with sharp, pinelike needles. The seeds or fruit of the tree, like its leaves, are strongly aromatic, almost perfectly round, green initially, turning frosted blue by spring. The high altitude types are flattened shrubs, growing out in a rough circle until, in older plants, they may attain a circumference of ten or fifteen feet. The needle leaves are quite sharp and prickly with a bluish green color, somewhat lighter underneath. The large purple berries are green the first fall and are intermingled with mature berries from the previous year, generally clustered on the underside of the outer branches.

HABITAT: The tree Junipers are found at lower altitudes in dry foothills, from 1,500 feet in southern California to 8,000 feet in the Rockies. Sometimes forming pure stands, they are frequently found in an almost symbiotic relationship with the Piñon Pine, particularly in Arizona and New Mexico. The high altitude Junipers (*J. communus, J. sibirica,* and *J. montana*) are generally found above 8,000 feet to about 10,000 feet, with *J. sibirica* growing up to timberline.

COLLECTING: The berries when ripe (bluish or purple), the leaves whenever needed. Berries, Method B; leaves, Method A.

PREPARATION: Berries or leaves, standard infusion, 2 to 4 oz. Tincture Method B, 1:5, 60% alcohol, 20 to 40 drops.

MEDICINAL USES: Primarily a urinary tract herb, most frequently used for cystitis and urethritis. The berries are the most effective. Use a teaspoon of crushed berries or a rounded teaspoon of the leaves, steeped in a covered cup of water for fifteen minutes and drunk, one to three cups a day. More effective and less irritating when combined with Manzanita or Hollyhock. Juniper should not be used when there is a kidney infection or chronic kidney weakness, as the oils are excreted in the urine and can be uselessly irritating to such inflammations. Munching a few berries an hour or so before meals will stimulate stomach secretions, i.e., hydrochloric acid and pepsin. The aromatic properties of all parts of Juniper plants have been used against bad magic, plague, and various negative influences in so many cultures, from the Letts to the Chinese to the Pueblo Indians, that there would seem to be some validity to considering the scent as beneficial in general to the human predicament. Overlapping traditions are useful in triangulating valid functions in folk medicine. If unrelated traditions say that Yarrow clots blood, it is easy to admit that such is probably the case; if they say that Juniper clears "bad vibes," many of us will back off and start to twitch skeptically. Our mechanistic approach to "primitivism" is too selective, accepting the possibility of drug effect on the one hand and nervously rejecting something as "subjective" as the warding off of bad influences on the other. In most non-Western peoples the two go hand in hand. A traditional Chinese herbalist may prescribe tiger teeth for impotence, knowing full well that the patient is suffering from cross purposes and simply wants a talisman to help in realigning internal disagreement. A Western professional in mental and emotional "sciences" is not supposed to rely on such nonsense and has to work through the patient's intellect, the same intellect that is probably the main cause for the impotence. In any respect, the Juniper berries, dry or moistened, can be thrown on hot rocks in saunas, sweat lodges, and the like, and the dried crushed leaves can be used as an incense. The leaves are traditionally carried about in pouches and clothes, often the only protection or medicine carried by a Tewa Indian. Consumption of the berries or leaves is not recommended during pregnancy; the volatile oils can have a vasodilating effect on the uterine lining.

CULTIVATION: The high mountain varieties are nearly impossible to bring down to sensible altitudes and the tree Junipers are best started from nursery stock.

OTHER USES: The berries are a necessity in venison marinades and in cooking any wild-tasting meat, from bear to *cabrito*. Ten berries per pound of meat is a good rule of thumb. They are also used in making sauerkraut and German potato salad. The leaves make a good garnish for fish and wild fowl, placed with the food shortly before removing from the heat.

MALLOW

Malva neglecta and others   Malvaceae

OTHER NAMES: Dwarf Mallow, Low Mallow, Malva, Cheese Plant, Cheeses

APPEARANCE: This common weed is a round-leafed, long-stemmed, ground-hugging mat. In rich, moist areas it may attain a foot or more in height, long stemmed and bright green; in less comely locations it resembles a scraggly, ill-fed garden geranium. The leaves are round, five- to seven-lobed and blunted palmate; the leaf stems are much longer than the leaves and spring at all angles from the main stems. The flowers are small, five petaled, cup shaped, from blue to white in color, clustered demurely along the larger central stems. The fruit is round, flattened, and resembles the wheels of cheese one so seldom encounters in supermarkets anymore. The root is light tan and rather deep. The whole plant has a slimy, mucilaginous sap when crushed between the fingers.

HABITAT: A common urban and suburban weed in all our areas, most frequent in the spring in our cities and towns, common on through late summer rains, in most farming and cattle lands and frequently planted or allowed its biennial freedom in orchards. The big Bull Mallow (*Malva borealis*) is less universal in habitat, mostly found in the farming areas of Utah, city lots of Arizona, and disturbed earth in coastal California. It may reach three feet in height, a virulent, macro-leaved version of the smaller plants. The Mallows are liked by chickens (sensibly enough) and are plowed under when possible, at least in those increasingly rare localities where chickens are allowed to "graze" on their own. Mallow turns the whites of their eggs pink. From sea level to grazing pastures at 8,000 feet, a weed by any standards.

CONSTITUENTS: Oleic, palmitic, and stearic acids, phytosterols, arabinose.

COLLECTING: The whole green plant, bundled and dried, Method A, chopped into small sections for storing. The deep roots of spring second-year plants may be used the same way as Hollyhock roots; same effects, just distinctly feebler.

It is advisable to wash the foliage well before drying, since its habitats and the downy surface of the leaves and stems combine to make it rather dusty. Mallow is a common "marker" plant, so avoid picking in areas obviously frequented by dogs.

PREPARATION AND MEDICINAL USES: Primarily used as a demulcent or emollient. The fresh or dried leaves make a soothing poultice, lessening pain and relieving and reducing inflammation. The tea is pleasant and green tasting, a good (if slightly slimy) beverage anytime but especially soothing to sore throats and tonsillitis. A rounded teaspoon to a tablespoon of the herb is steeped in water and sipped slowly as often as needed. For somewhat more extended use (a sore throat that has continued for a few days, for example), make a standard cold infusion in the morning and sip up to a quart of the tea during the day. The tea (hot or cold) will help indigestion and stomach sensitivity, and the quart-a-day cold infusion regimen helps obstinate bladder and urethral irritability, especially when it results from hospital catheterization. The tea is traditionally drunk in New Mexico to facilitate labor in childbirth and as a neonatal wash for skin irritations.

Up until the 1930s it was a part of the official *Species Emollientes,* N.F. an accepted poultice mixture for breast and soft-tissue dressing. See Formula #9 (page 138).

The best way to use the poultice is to mix the herbs and water, heat slowly over a flame until it bubbles like oatmeal, enclose it in several folds of lightly oiled cheesecloth, press it hot over the engorged area, press out until one-half to one inch thick, leave on for up to an hour. Cover with freshly moistened hot towels as the poultice itself cools down. This rather archaic procedure still works, by the way; medical practice has changed, not the human body. The poultice helps break down disorganized tissue, transudates and exudates, and speeds up the resolution of the inflammation by white blood cells and fluid movement.

MANZANITA

Arctostaphylos spp.   Ericaceae
OTHER NAMES: Madrono Borracho,
Pinguica

APPEARANCE: The most distinguishing characteristics of the various Manzanitas are the smooth, matte-finished red bark, the twisted branches, and the large, oval, leathery leaves. Only in the oldest branches and trunks does the bark shred. The heartwood of the large branches and trunks is usually about the same shade of red as the bark. The leaves are round-oval to pointed-oval, dull green, smooth and thick as leather. The flowers bloom from February to June, forming little nodding clusters, pinkish and urn shaped, maturing into tart, red, mealy berries containing from four to ten seeds. Manzanita varies in height from three-foot-tall mat shrubs to thirty-foot-tall individuals. They can be confused with the related Madrone (a tall, open tree with larger leaves but identical in use), Jojoba (greenish bark, lower and hotter elevations), and Silk Tassel (pronounced central leaf vein, intensely bitter taste).

HABITAT: Although traditionally associated with California and the coastal mountains, Manzanitas grow as far east as the Guadalupe Mountains of New Mexico and the Davis Mountains of Texas, and nearly all points between. It is fond of gravelly hillsides, acidic post-climax old forests, and the canyons that drain down from the Ponderosa/Piñon interfaces. From the Mojave Desert and Panamint Range east, the Manzanita species are more shrubby and thicket-forming than the taller species found in California. Manzanitas grow from 7,000 feet in the dry foothills of the Jemez Mountains of central New Mexico to nearly sea level around California's Death Valley, Imperial Valley, and the Pacific Coast.

CONSTITUENTS: Arbutin, hydroquinone, methyl-arbutin, ericolin, and 15–25% tannins by dry weight.

COLLECTING: The berries when just ripened and not yet mealy (April or May), dried Method B. The leaves collected at any time, although somewhat stronger just after flowering; dry, Method A, stripping the leaves off the stems when dry.

PREPARATION: The leaves, chopped or ground for use, two to four ounces of the Standard Infusion, three times a day. Tincture Method B, 1:5, 50% alcohol, one to two teaspoons in water, three times a day. Sitz bath, 1 pint Standard Infusion (1:32) per gallon of water.

MEDICINAL USES: Used in nearly the identical way that its close north-country relative Uva-Ursi (*Arctostaphylos uva-ursi;* see *MPOMW*) is, the Manzanitas' constituents, particularly arbutin and ericolin, break down into various quinones in the urine, giving the tea or tincture the disinfecting qualities that are so useful in mild urinary-tract infections, bladder gravel, and the irritations that follow catheterization. For acute cystitis and urethritis that follows sugar, grain,

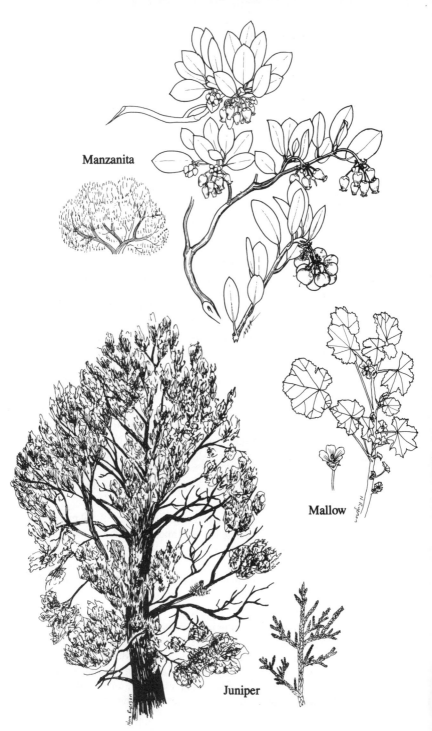

Manzanita

Mallow

Juniper

starch, or fruit binges with short-term alkaline urine (I include fasting, fruit-juice fasts, and raw-food "cleansing" under the general category of binging), use the tea or tincture as recommended, alternating with cranberry juice. If the irritation occurs from overly acidic urine (short-term excesses in meat, cheese, and milk products, steroid therapy, just recovering from a GI infection, heavy Premenstrual Syndrome), combine the Manzanita with Hollyhock, Malva, or Puncture Vine, and drink lots of Alfalfa or Red Clover tea (see *MPOMW*). Manzanita is for short-term use, at least internally; the tannins can irritate the stomach lining and liver with long-term use. If it doesn't help in two to three days, do something else.

The sitz bath (I have been told) is nearly miraculous for the first week after labor, or for getting over the discomfort of rectal or uterine surgery. Sit in it for an hour, once or twice a day; make it fresh each time, nursing or reading a mystery (or an herb book, for that matter) while you do it. For vaginitis, a standard infusion (1:32) can be used once a day for up to a week as a douche.

CONTRAINDICATIONS: The vasoconstricting effects as a tea make it inappropriate for use in pregnancy.

OTHER USES: The berries make a pleasant, tart jelly, best combined with apple juice. Leaves are widely used for smoking, either with tobacco, other herbs, or alone. The dense root crowns formed by the fire-resistant Manzanitas have been used for carving smoking pipes, like briar. Some of the bushes or trees have sufficiently thick wood for carving and woodworking. The grain varies from deep yellow to brick red and polishes beautifully.

MARSH FLEABANE

Pluchea camphorata   Compositae

OTHER NAMES: *P. purpurescans,* Camphorweed, Stinkweed, Salt Marsh Flea-bane, Canelón, Santa Maria

APPEARANCE: Two- to four-foot-tall annuals or (in parts of our area) short-lived perennials; they have bright pink to magenta flower tops that bloom from early summer to mid-fall. The leaves are alternate, with ovate or lanceolate, toothed blades, somewhat stem-clasping, and sort of sub-sticky. The whole plant smells strongly of slightly rancid camphor or mothballs. With its many bright, genial upright flowering stems and overall demeanor, it looks sort of like Yarrow or Boneset or even a tame thistle. If you do much gathering of plants for tea or medicine, or even grow them, chances are that if you see a colony of these plants along a ditch or stream you will say to yourself, "I wonder what this is and what it's good for?" Of course, that's why you have this book in your hand, isn't it.

HABITAT: From southern California to Florida and north to Maine, in wet places with brackish or salty water. Marsh Fleabane is abundant but never predictable. It doesn't like Nouveau Wet places, just Old Damp. It isn't encountered very often growing around Corps of Engineer diversion dams, aqueducts, or irrigation projects, but instead prefers those old original marshes that the ducks and grebes prefer.

I don't know if there is any real difference between *Pluchea camphorata* or *Pluchea purpurescans;* I have observed these plants growing from the Sacramento delta to the Ouachita Mountains in Arkansas and it's all the same plant to me.

CONSTITUENTS: Cuauhtemone, artemetin, herbacetin, quercetin, and various eudesmane, phthalic acid, and carvotagetone derivatives.

COLLECTING: Bundled and dried when flowering, Method A.

STABILITY: As long as the plant retains its characteristic odor, usually for one to one and one-half years.

PREPARATION: Standard Infusion, two- to four-ounce doses, preferably hot, except for an eyewash; for that purpose make the tea at the same strength (1:32) from an isotonic water (one-half teaspoon of salt dissolved in a pint of water, cooled to body temperature).

MEDICINAL USES: The hot tea will predictably stimulate perspiration, much in the same manner as Pleurisy Root (see *MPOMW*) or Pennyroyal, with an increase both in the liquid and waste products. Without an elevated temperature or increased metabolic rate, the main effects of the tea are to increase the urine, also both in liquid and waste products.

It is a safe and reliable menstrual stimulant when the flow begins late, is scanty, and there are clotty cramps. It, like some others of our herbs, stimulates endometrial mucus secretions, those that help to facilitate menses by inhibiting clotting and microbes. Further, Marsh Fleabane is distinctly antispasmodic, thus aiding the cramps. It also will inhibit the spasms and cramps from both diarrhea and just plain old stomachache.

The eyewash helps reduce redness and pain from simple hay fever, wind and dust. Drink the tea and apply it externally. As with all herbal eyewashes, it is important to make a fresh tea each time you bathe the eyes, and to make it with isotonic water. Years ago, when most eyewashes were prepared for the patient by pharmacists, slight irregularities in the formulations would bring about the introduction of bacteria into the eye from the pharmaceutical. If *they* couldn't do it right sometimes, *we* shouldn't try; fresh each time.

The tea is decidedly a stimulant, much as is camphor; a concentrate was once marketed in the early years of this century as a substitute for that old "Devil Coffee." It is not recommended for people who get migraine headaches, however, because, unlike coffee (a vasoconstrictor), Marsh Fleabane is a vasodilator.

CONTRAINDICATIONS: Due to its stimulation of the uterine mucosa it is not recommended during pregnancy.

MATARIQUE

Cacalia decomposita   Compositae
OTHER NAMES: Maturique, Maturín, Indian
Plantain, Buffalo Root

APPEARANCE: This is a large, handsome plant, forming lively, filigreed, basal leaves finely divided several times until the rosette resembles a long-stemmed mass of parsley lace. It sends up a three-foot-tall flowering stalk in the late summer that blossoms into numerous dense flower heads, with each of the cream-white flowers extending well past the bracts. The blooming Matarique resembles, at this stage, a cross between a big Yarrow (see *MPOMW*) and Celery blooms. These mature into short, dandelion-like, wind-carried seeds. The root is hairy and woolly at the junction of the plant's stalk, the rootlets forming a shallow, broad, octopoid, fleshy mass, oozing yellow-orange sap where broken.

HABITAT: Found on rich gravelly slopes, from 5,000 to 8,000 feet, in the desert mountains of southern Arizona, southwestern New Mexico, and all the northern ranges of Sonora and northwest Chihuahua. If the range is wet enough to support a stand of Chihuahua or Apache Pines, start looking for Matarique on the leeward and shady slopes and canyons above 5,000 feet. It is not a terribly common plant but is well established in some of our ranges and recuperates well from "culling."

As of this writing, I have gathered the same saddle in the Chiricahuas for nine years, either for commercial use or with students. There are now more Matariques than when I first started. The mature seeds gathered in the fall germinate well in flats and are such attractive plants that they would grace any garden.

CONSTITUENTS: Cacalol and cacalone (furotetralin sesquiterpenes), decompositin (an eremophylane sesquiterpene), maturinone (a quinone sesquiterpene), and chastacine (a pyrrolisidic alkaloid).

COLLECTING: The roots, gathered from midsummer to late fall, dried, Method B.

STABILITY: The dried, whole root is stable for at least two years; if ground or chopped initially, less than a year.

PREPARATION: The dried root tinctured, Method B, 1:5, 60% alcohol. The root tea as a cold Standard Infusion, one to two ounces. The capsules are more irritating than the water or hydroalcoholic extracts, and so are not appropriate.

MEDICINAL USE: *External.* The tincture or tea works in a way similar to and, according to Martinez, more effective than Arnica, applied as a liniment and allowed to dry. A salve, made Method A, can also be used. It is used for sprains, bruises, and joint pain, with or without inflammation. Specific symptoms are pain aggravated by movement, with little pain at rest. It stimulates the reabsorption into surrounding tissues of the congested fluids (transudates and exudates) that cause the engorgement and pain. Since true Arnica doesn't grow in most of our area (unless you live in the western Great Basin or northern New Mexico), this plant, along with Camphor Weed, fills an important gap for those of you wishing to gather your own plant medicines.

Internal. Matarique is a strong antispasmodic for cramps and tenesmus of the intestinal tract and bile ducts, but its striking hypoglycemic effects make it seldom useful for this purpose.

It is one of the most widely used diabetic herbs in our area, used from Mexico City to the Navajo Nation. It should be considered as an early treatment for adult-onset insulin-resistant diabetes. In this acquired (not inherited) disorder, the traditional uses are for the overweight and sedentary diabetic. Clinical trials in Mexico in the 1930s indeed supported this Native American and New Mexican Spanish traditional use of Matarique; obesity and poor diet were the picture for the use of Matarique. Like Prodigiosa, it decreases the oversecretion of glucose by the liver. The exact mechanism is still unclear (as is much about hyperglycemia); presumably it blocks gluconeogenesis, rather than having any direct effect on the insulin-glucagon cycle.

I have found that using a cup or two a day of Prodigiosa (morning and afternoon) seems to be a good maintenance program, with Matarique being used for periods of acute glucose elevation. As in all forms of diabetes and simple (!) hyperglycemia of the middle-aged, it is important to decrease saturated and poor-quality fats, refined carbohydrates, and alcohol intake. Matarique also seems to stimulate liver protein metabolism and the quality of gastric function; so try 20–30 drops of the tincture afternoons and evenings before meals, together with a similar amount of Algerita (or Oregon Grape Root) tincture, initially for a week or two. This should have a subtle effect on food cravings and binging. You should feel less attraction to greasy and sweet stuff, more interest in good bread, whole grains, or more proteiny "heavy" foods. After a week or two, slip into the Prodigiosa tea and then reserve the Matarique for special, more acute uses.

In the tradition of early plant experimenters and homeopaths, I "proved" the herb with two successive classes of helpless students a few years ago. We all took 30 drops of the root tincture, each time around 10:30 in the morning. Those several that suffered from allergy-related rebound hyperglycemia (eat a big meal, fall asleep an hour later, wake up with a stuffy nose) got either frontal headaches or got dizzy and groggy. The rest of us got the stoned-out munchies and raided the cafeteria en masse for *anything* that was available and highly caloric. So it goes.

CONTRAINDICATIONS: Juvenile or insulin-dependent diabetics; pregnancy; liver or renal disease, stomach ulcers, insulin users.

MESQUITE

Prosopis julifera, P. pubescens
Leguminosae

OTHER NAMES: Honey Mesquite (*P. glandulosa*), Screwbean, Tornillo (*P. pubescens*)

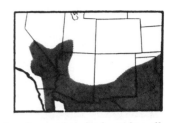

APPEARANCE: The Mesquites (including Screwbean) are spiny shrubs and small-ish trees. Many of us call any desert or dry-grassland shrub Mesquite, but the differences are many and, for medicinal purposes, important. When in flower in the spring, the Mesquites have a long, yellow flower spike from one and one-half to three inches long; each flower in it has ten stamens. Other legume shrubs that might be confused with it have twenty or more stamens, and most of them form balls, not spikes. The thorns of the Mesquites are usually paired and straight. *Prosopis pubescens* (Screwbean) has the spines emerging from the branches along with the leaf stems. The several varieties of *Prosopis julifera* (Mesquite) have the spines emerging from the branches and are distant and unrelated to the leaves. Several of our Mimosas may confuse the issue, but their spines are curved back like those of a rosebush and their pods are usually spined as well; Mesquite thorns are always straight. Finally, the long straight pods of the Mesquites do not open when ripe, and in fact may cover the ground in the fall and winter still unsplit. Other legumes of similar appearance have pods that split open and cast their seeds passively; Mesquites prefer to be eaten and excreted by range animals. The pods of Screwbean look like little . . . screw-beans; like the other Mesquites, they too do not split. Medicinally both species are interchangeable; so I will refer to them all henceforth (including the four recognized but freely interbreeding varieties of *P. julifera*) as Mesquite.

HABITAT: The most common shrub or small tree in all of the Southwest. It fills most arid and semiarid lands above Chaparral (*Larrea*) and below the Juni-per/Piñon belt, spilling into both areas from above and below. Altitudes may vary from 1,000 to 5,500 feet, but take my word for it; if you are in any part of the Colorado or Mojave deserts or the deserts of Sonora or Chihuahua, you are not very far from some Mesquite. If, further, you are in dry areas of Texas, Oklahoma, Arkansas, or even Kansas, you *still* aren't too far from some. Only in the Great Basin are you (sort of) out of luck.

Much of this invasion of the genus into former grasslands (where it did not formerly grow) is the direct result of overgrazing a century ago, with the diminishing of native perennial grasses and the spread, by cattle droppings, of Mesquite seeds. Apaches, those wild mountain folks who used to share this land with us, are indirectly responsible for the extensive *Prosopis* presence in much of Arizona, New Mexico, and West Texas. The Military-Industrial Complex of the 1865 to 1890 period in the Southwest (as many as 100,000 military personnel and family and horses were stationed in forts and communities to combat per-haps as many as 3,000 armed Apaches [only 100 by 1885]) had to be quartered and fed. This resulted in overgrazing and land abuse from the private sector as well, a lumbering, self-perpetuating, low-tech version of present-day defense industry boondoggling. If the Apaches hadn't resisted so well, I can almost imagine someone inventing an equivalent, if only to "open up" and Civilize the area. When Geronimo surrendered, and the military presence trickled on back east (or stayed when mustered out), much of the economy of the Arizona and New Mexico territories took a nosedive. The Mesquite had by now taken over much of the old grasslands.

CONSTITUENTS: Leaves, pods, bark: 5-hydroxytryptamine, tryptamine, tyramine, prosopine; gum: L-arabinose and D-glucuronic acid.

COLLECTING AND PREPARATION: The leaves, flowers, pods, and bark, whenever needed or available, dried however is appropriate. The bark and smaller branches should be cut into smaller sections when fresh, so they can be more easily powdered when dry. The pods are used whole for decoctions or syrup. The gum is collected carefully, with as little adhering junk as possible; in dry summer months some resin will be found below the bush on the ground. Many areas have gentle weather (for Mesquite) and the gum is seldom encountered; so if you live near or can visit regularly a large stand of them, you can urge them to form gum. In the spring or early summer, walk around a patch of bushes and tear off some lower live branches from the main trunks. When you return in three or four weeks the plants will have healed the scars and usually secreted gum along the edges of the old wounds.

PREPARATION: For a tea or cleansing disinfectant wash, make a Strong Decoction of the leaves, twigs, bark, and/or pods. Internally, take 2–3 ounces every four hours; externally, use as needed. For an eyewash, put 5–6 well washed pods somewhat crushed into a pint of isotonic boiling water (½ teaspoon of salt to a pint of water), steep until body temperature, and strain. Apply to the eyes and forehead; make fresh each time. To make Mesquite mucilage (identical in function to Gum Arabic mucilage) take 1 part by weight of clean Mesquite gum tears, wash briefly in cold water, and drain; place them in enough warm distilled water so that both the tears and the water equal three parts by volume of what the tears weighed. Place in a stoppered bottle, shake occasionally until the gum has dissolved in the water. Strain and refrigerate; this will be stable for a week or two. It can be stabilized for much longer periods by adding 1 part glycerin to 10 parts mucilage by volume. If the mucilage starts to smell vinegary, discard it.

MEDICINAL USES: Internally, the tea of the powdered plant is strongly astringent and rather antimicrobial; alone it will inhibit diarrhea and other GI tract inflammations, from dyspepsia and ulcers to colitis and hemorrhoids. Add it to Desert Willow, Tronadora, Trumpet Creeper, Chaparro Amargosa, Silk Tassel, or Echinacea—whatever you have gathered—to make a formulation to benefit other types of intestinal distress, from shigellosis to amebic dysentery to food poisoning. See Formula #10 (page 138).

External, the tea is a splendid cleansing wash for most any broken-skin injuries. If there is bruising or edema involved, alternate the wash with fillets of Prickly Pear.

The pods, made as an eyewash, help conjunctivitis of any type and work

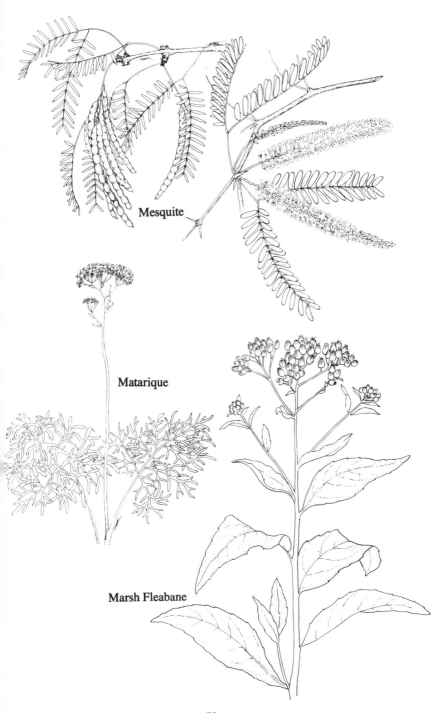

Mesquite

Matarique

Marsh Fleabane

nicely for the pinkeye of children, household pets, and livestock.

The mucilage, a thick, slimy goo, is guaranteed to soothe red and painful membranes in sore throats, laryngitis, stomach inflammation, and acute pain. For peptic ulcers it has the advantage over the usual chalky stuff, as it will not affect digestion and nutrient absorption from the small intestine and can be taken for long periods of time. Alone or combined with Fremontia (see *MPOMW*) it may be used as a basic therapy to restore intestinal mucosa health in more serious GI diseases, or in recuperating from intestinal surgery. Doses of the mucilage are one to two tablespoons, as frequently as once an hour if needed.

OTHER USES: The pods are a widely used food, and the methods of preparation are recounted extensively in a number of ethnobotanical texts. The sweet pods also make an excellent molasses, not recounted in the texts. With a ratio of ¼ pound of washed pods per quart of water, cook them for at least twelve hours at the low setting in a crock pot; strain, and reduce the tea by slow boiling to a thin, syrupy consistency. When cooled, it thickens into a strong, robust, sweet syrup. Collect the pods in September and October, culling and discarding those that are light and hollow from insect predation, keeping the heavier, full pods for making the molasses.

MILK THISTLE

Silybum marianum   Compositae

OTHER NAMES: *Carduus marianum, Cnicus marianum,* Variegated Thistle, St. Mary's Thistle

APPEARANCE: Milk Thistle is a tall, coarse, dangerous-looking annual or biennial, usually standing four to six feet tall by summer. The flowers extend well above the lower leaves, the stems clasped by smaller, lobed leaves. The flowers themselves are of moderate size, one per stem, pink to light purple. These moderate, typical thistle flowers are ringed, like a bulldog with a spiked collar, by thick, terribly sharp, spiny bracts that extend the diameter of the head by double, pointing in all directions. Considering the spines that cover the rest of the plant, this is a case of floral overkill. The lower stem leaves and basal leaves may be two feet long, undulant, spiny, and strongly marked by white blotches and mottling. The leaves are wavy and deeply cut all along their edges, and almost every part of the plant is covered in half-inch-long yellow spines.

The seeds, usually mature by late summer and pickable until rather late in the year, are a quarter-inch long, flattened, and shiny mottled-brown. Thoroughly mixed up with the ripe seeds (properly called achenes) are a lot of chaffy and prickly hairs (properly called pappus hairs, improperly called bad names).

HABITAT: Abandoned fields, old pastures, alongside irrigation ditches and waterways of the farmlands of California; Arizona (less common) from Phoenix south; New Mexico's Rio Grande Valley, east and north. Although a common weed in many parts of the world, especially in temperate zones, plants growing in the dog latitudes (that's us) have more potent seeds than those from the north. Therefore, considering its value as a medicine, you might feel us to be fortunate . . . until you try to process it! The Great Basin has little Milk Thistle, except around the Great Salt Lake and parts of Idaho. Except for coastal California, it is sporadic in distribution, and it might be easier to check with your local agriculture extension service. *They* know the plant very well. It plants well from seeds; so when you get some, plant them in an appropriate place (don't tell anyone . . . they might not appreciate it) and harvest it the next year.

CONSTITUENTS: Silymarin (a complex), flavolignans, silybin, silydianin, silychristin, taxifolin.

COLLECTING: (gulp) Remove the mature flower heads from the receptacle. You may have to invent your own procedures, as there is no easy way. The big, spiked collar can be removed with shears and the remaining head run loosely through a hand mill or quickly in a blender to separate the seeds from the head; winnow in front of a side-placed fan or trickle down a large box top or tray. However it gets done, the seeds need to be fairly well separated from most of the attached pappus hairs. After all, Milk Thistle is not an herb to be a perfectionist about. That way lies madness.

STABILITY: The seeds are stable for at least a year; so don't worry.

PREPARATION: The seeds should be dried for a week in flats, then grind them. A coffee mill seems to work better than a blender. Milk Thistle may be taken in capsules (two #00 capsules at a time) or made into a tincture, Method B, 1:5, 50% alcohol. Take frequent doses for acute conditions, one or two a day for chronic conditions; capsules as above, tincture ¼ teaspoon. Too much can cause a slight dizziness, but that means a lot too much.

MEDICINAL USES: Milk Thistle, in the form of Silymarin complex, is widely used in Europe, clinically, experimentally, pharmaceutically, and over-the-counter for self-medication. Its present role as an accepted liver protectant and immunologic support is quite recent, and the monographs and drug brochures from there talk about its "new" uses. In actuality, a monograph by a Dr. Lobach in the *American Journal of Medical Science*, April 1859, relates its successful use in hemorrhagic blood disorders resulting from liver and spleen disease, as well as its beneficial effects on liver congestion and portal hypertension. Ellingwood, Felter, and Lloyd wrote extensively about its medical uses for blood dyscrasias, hepatosplenamegaly, and portal congestion at the turn of this century.

One of the reasons we probably won't see some of these European pharmaceuticals made from Milk Thistle in the United States, at least in standard practice medicine, is that it—like so much that is valid in the center of herbal medicine, acupuncture, naturopathy, and alternative medicine generally—does not treat a disease. It strengthens functions and processes of metabolism, it doesn't treat anything specific.

When function breaks down, you have organic disease; something is broken, it won't work, the fleshy vessel is cracking, you probably need medical attention. Although most organic disease arises from functional imbalances, these are difficult to diagnose and treat, and some physicians even view the very idea of subclinical, functional pathologies to be hogwash. To them, when something breaks, you have a disease; even if the problem is obvious, the medical approaches are usually too strong to help.

Stated another way, a functional disease is like leaving the thermostat up to high on your central heater. If the heat is on all the time, the energy costs can get staggering and either the heater burns out from such constant, unrelenting use or the gas company turns it off when you can't pay the bill. In both cases the functional has become the organic; it's broken or you can't afford to pay the bill because *you* are broke. Milk Thistle helps to lower the thermostat, clean the vents, blow off the junk on the air filters, etc. It doesn't help if the furnace is broken. For a broken furnace you need a repair person (physician). In this country we have no heating system maintenance specialist in medicine, nobody to tell you how to turn down the thermostats. Before I get organically tangled in my similes and over-functioning metaphors, let me get on with what Milk Thistle *does* and leave the evaluation to your intelligence.

First of all, the liver. It decreases the damage to that organ from alcohol, both short-term toxicity and long-term degeneration. It helps normalize the ability of the liver to make phospholipids, including cholesterols, either for the current drinker or for the recovering alcoholic. Consider that in order to make a cell or repair a cell wall you need good proteins and good fats (phospholipids) to make the membranes; the liver does most of the work in supplying the body with these building materials. Half of the cholesterols in the body are used in insulating the brain. As that organ shrinks in heavy drinkers, it misfires and sends crossed signals. Milk Thistle helps our alcoholics recover better and more intact.

In general, the liver specializes in breaking down waste products brought to it in the blood and building useful pieces back up from the ashes. Most cells can do this, but a healthy liver does it for them, leaving their energy to be expended on *their* specialized functions, so the liver has more energy for *its* specialized functions, on and on. Only the liver, however, can make the blood work; only it can clean out the metabolic wastes that come back to it. Only the liver can maintain the healthy colloidal osmosis from its protein synthesis, and the liver alone can turn active fuel into stored fuel and back again, at least in the quantities that the muscles and brain need them. In light of this, it's useful to know that Milk Thistle is ten times more effective than Vitamin E in inhibiting the lipid peroxide buildup inside those hyperactive liver cells, preventing the damages from these free radicals, stimulating as well the synthesis of RNA in the hepatocytes of the liver, and enabling them to make more of the enzymes that break down and build up the blood constituents. It also protects against, to some degree, the damages of recently ingested heavy metals.

In some environmental stress, with excess anabolic response, Milk Thistle increases the liver's synthesis of unsaturated fatty acids and high density lipids, thereby helping to prevent placquing and hardening of the arteries. Take your basic, hot, vibrant, sthenic individuals as an example—warm-skinned, moist, staying up late, getting up early, borrowing life from their next decade, candle-at-both-ends people, burning bright but short. Milk Thistle helps cool them down, decreases the damages from their high thermostat, cleans out the vents . . . here go the metaphors again.

When the liver is overworked, or unequal to the task our brain gives it, its blood vessels enlarge, the fluids move more slowly through it as it tries to increase its working area; it gets enlarged and congested. Blood trying to get in backs up; this is called portal hypertension. The portal vein (really a misplaced artery) draws blood from the intestinal tract and the spleen; if the blood can't get in fast enough, the blood drawn from closer tissues get in first and the other blood backs up. The farther from the liver, the more the backup. In mild portal hypertension, blood from the colon and pelvis can bypass through smaller veins into the general circulation. As these veins aren't large enough, the returning blood from the legs gets backed up going into *them*. With all that used junk-blood from the colon going through them, they enlarge, balloon out, and become venosities and vericosities. And so it goes, with pelvic congestion, hemorrhoids, hydroceles, chronic urethritis, enlarged prostate and cervix, eventually with leg vericosities. Milk Thistle helps.

Further (I'm almost through), the ability of the blood to carry nutrients, fuel, and building materials out and waste products in is largely dependent on their finding proteins and corpuscles to cling to. Milk Thistle, by subtly improving the quality of blood proteins and the quality of red-blood-cell surfaces, helps move things around through our primeval mother ocean, the blood.

Generally, Milk Thistle combines well with Echinacea, Yerba Mansa, and Cypress in overt infections; with Ocotillo and Indian Root for fluid retention and pelvic congestion; and with Red Root in lymphadenitis and such infections as Epstein-Barr Virus and cytomegalovirus; it has shown clinical value in aiding the Crohns Disease patient as well.

But mostly it helps you work better so you don't get sick.

NIGHT BLOOMING CEREUS

Peniocereus (Cereus) greggii
Cactaceae
OTHER NAMES: Queen of the Night, Reina de la Noche

APPEARANCE: This cactus sends out beautiful two- to three-inch white flowers during June; each blooms at night and fills the air with honeysuckle-like fragrance. Usually they all bloom in one or two nights, but sometimes a large individual may apportion them out over a week. By fall they mature into red, pear-shaped, delicious fruit. The rest of the time, the excitement over, this cactus sits in the middle of Chaparral, Mesquite, or Palo Verdes, undistinguished and dead-green colored, formed of tall skinny stems one-half to one inch around, two to six feet tall, weakly spined along the four to six ridges of the stems. These arise out of a huge, turnip-shaped brown tuber that may weigh as much as ninety pounds, although five to ten pounds is the norm.

HABITAT: Coarse sand and gravel plains and bajadas, from 1,500 to 4,000 feet. As mentioned, most of them, with their great height and slender stems, need the protection and support of shrubs and trees to grow within or against. Although seldom noticed, they are really rather abundant, growing from nearby Kingman, Arizona down through southeastern Arizona, southwestern New Mexico, West Texas, and northern Mexico.

CONSTITUENTS: Peniocerol, viperidone, desoxyviperidone, viperidinone, B-sitosterol, and (probably) caffeine.

COLLECTING AND PREPARATION: The root is the strongest part of the plant, but the renewable stems serve the same function, though more feebly. If you just remove the stems, the tuber will regenerate the green stems the same or the following year. Tincture the stems fresh, Method A. If you are surrounded by the plants, take a part of one of the roots as well as the stem . . . generally, just taking the green part seems better to me; they're such nice plants. The dose of the tincture is ¼ – ½ teaspoon two times a day.

MEDICINAL USES: Called "Pain in the Heart" by Death Valley Shoshones, the root and the stems are a useful cardiac stimulant. It is not a digitalis-like cardiotonic. Instead, it is useful in helping the tachycardia, arrhythmias, and vague chest pain and shortness of breath often associated with tobacco and caffeine abuse, and for those people that get adrenalin rushes with a panicky tightness in the chest and intercostal pain. These are not necessarily any more than a coronary equivalent of ulcers or hay fever.

If the doc says it's just nerves (which you already knew), and all that is offered is some form of tranquilizer (which you don't want), try some of this strange little cactus, and maybe some Passion Flower tea . . . and cut down on those recreational drugs (like coffee or tobacco) and take 350–500 milligrams of L-tryptophan in the evenings for a couple of weeks. You're probably secretly afraid you have a bad ticker (which adds to the stress), so bite the bullet, get a diagnostic checkup, *then* (if needed) take the herb. Ignorance is *not* bliss in such matters.

The root tea or stem tincture with a little Wild Cherry bark (½ teaspoon) is a good first stage treatment for incipient bronchitis, with rapid, shallow breathing and a dry, tight sensation across the chest.

Several diverse groups of Native Americans use the stem to treat diabetes; I personally have no clear idea about how well it works, but, considering the results of tests with Prickly Pear, it may have promise. Of course, Prickly Pear is much more widespread. One ethnobotanical text alleges that the Apaches used the tea as an intoxicant; another expert said that was nonsense. I don't know how well it works, either. Organoleptic provings (that means *I* get to try it out) on such alleged intoxicants usually put me to sleep, give me a headache, or make me throw up . . . I have better things to do these days.

OCOTILLO

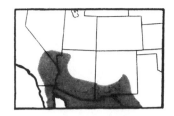

Fouquieria splendens   Fouquieriaceae
OTHER NAMES: Candlewood, Couchwhip

APPEARANCE: There is little chance of mistaking the Ocotillo, since most of the year it is a six- to twenty-foot-tall mass of many viciously spined stems radiating from a central, shallow root system. Standing in a field of these spiny-whipped plants is like being in the middle of a heavy-metal nightmare. In the spring (April or May) they bloom in scarlet fury, hundreds of bright tubular flowers crowding the ends of the branches. After spring or summer rains, the leafless branches burst into leaves; the rest of the year the photosynthesis occurs under the lacquery bark of the stems. These sturdy, unique plants never seem to die of natural causes; eventually their shallow roots, erosion, and wind combine to topple a giant here and there.

HABITAT: Ocotillo is found in all the deserts of the Southwest (see distribution map), from sea level (Imperial Valley, California) to over 5,000 feet (central New Mexico). It is found in great abundance around the edges of desert valleys, on the rocky hillsides, bajadas, and mesa tops.

CONSTITUENTS: At least 12 iridoid glucosides (most are galioside derivatives), monotropein methylester, adoxoside, loganin, and various polymerizing waxes and resins.

COLLECTING: With good gloves, of course! Larger plants will have older, thicker stems in the center with bark so deep that the thorns will be nearly or totally overgrown. These are easier to work with. A four- to six-foot section should be cut from one stem (more than enough medicine for a family) with limb shears and further cut into 6-inch sections. Work the outer bark off of the center wood core and discard the latter or use it for fire kindling.

PREPARATION: A fresh bark tincture is the only practical way to prepare Ocotillo. Chop or snip the freshly removed bark into ½-inch pieces, prepare tincture, Method A.

MEDICINAL USES: The tincture is taken in a little warm water every three or four hours, usually in doses of 25–35 drops. It is useful for those symptoms that arise from pelvic fluid congestion, both lymphatic and veinous. It is absorbed from the intestines into the mesenteric lymph system by way of the lacteals of the small-intestinal lining; this stimulates better visceral lymph drainage into the thoracic duct and improves dietary fat absorption into the lymph system. With fewer dietary lipids going into the liver by the portal blood, there is less tendency for the intestinal blood to back up (portal hypertension) and less stagnation in the pelvis and upper thighs. Most hemorrhoids are helped by Ocotillo, as are cervical varicosities and benign prostate enlargements. The same is true of a frequent need to urinate, with dull ache but no inflammation of the urethra, and

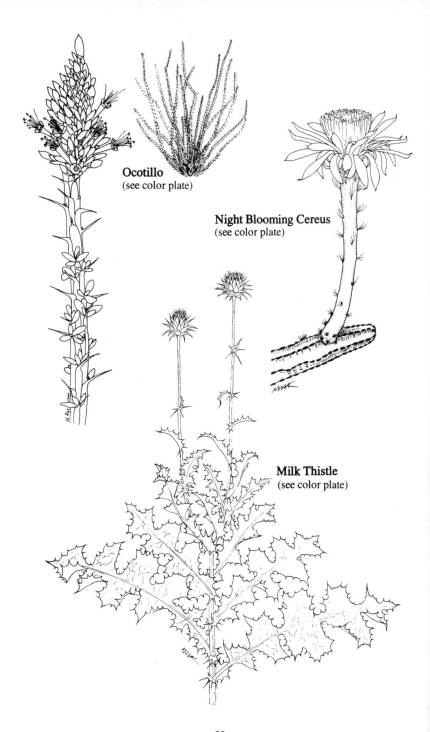

Ocotillo
(see color plate)

Night Blooming Cereus
(see color plate)

Milk Thistle
(see color plate)

82

the kinds of varicose veins and piles worsened by constipation or poor digestion. These are a variety of problems, often with other, more primary causes, but all of them are aggravated (or begun) by poor fluid movement and congestion in the lower viscera and pelvis.

The Cahuilla Indians of California used a strong tea of the root for painful, moist coughing in the aged; Apaches took baths in and drank the tea of the root for fatigue and swollen, tired limbs. The flowers, gathered in the spring, make a pleasant and elegant sweet/tart tea, either steeped fresh in the sun or dried, crushed, in hot water. The only problem is how to coax them off a twelve-foot-high, well-armed Ocotillo plant.

Note: Ocotillo is a protected plant in Arizona. Cutting a six-foot section from a 15-foot-tall plant with fifty branches in no way harms the plant, and it is found in abundance in five states and the whole northern half of Mexico; so use your judgment. If nothing else, get a section from someone's private property, or gather sections from uprooted plants at your nearest foothills site where they are bulldozing for a new subdivision.

PASSION FLOWER

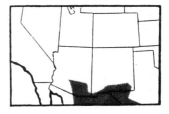

Passiflora mexicana, P. foetida, P. tenuiloba, etc.   Passifloraceae

OTHER NAMES: Maypops, Pasiflora, Passion-Fruit Vine, Itamo, Pasionaria

APPEARANCE: Passion Flowers are vines, period. The leaves are palmate and usually three-lobed, although *Passiflora mexicana* has but two and *P. tenuiloba* may have five. So, you say, does almost every vine around, from wild grape to Virginia Creeper . . . even Poison Ivy. What makes *all* the Passion Flowers unique is the flower. There are five sepals and five petals, usually greenish white or blah-beige, and normally the same length. There is a distinctive fringe of petal-like hairs (the perianth) that emerges from the calyx and is the most distinguishing flower part; usually they are pink, blue, or purple and are what strikes your eye at first. The pistil is stalked, ending in three stigmas projecting sideways from the summit. There are also five stems that extend out from between the perianth and the stigmas. Complicated but comely. Early botanists saw this all as representing the Passion; ten "petals" for the apostles (excluding Peter and Judas), three stigmas for the nails (or the Trinity), the stamens were the five wounds, the tendrils were the scourges, etc.

HABITAT: This can be a little confusing, as there are native Passion Flowers growing in moderate isolation in Arizona, New Mexico, and Texas (see map), and there are cultivated Passion Flowers covering the back alley fences of the older sections of our cities.

Passiflora mexicana (native) has pinkish flowers, two-lobed leaves, and grows in canyons of southeastern Arizona and Mexico; *P. foetida* (it smells rank, hence the name) has hairy grape-like leaves, lilac flowers, and is found in southern Arizona (uncommon) and the southern Rio Grande Valley and Edwards Plateau of Texas and Chihuahua; *P. tenuiloba* has thin, fingery leaves, bland, light-green-to-yellow flowers and is found in southeastern New Mexico and through the same area as *P. foetida*.

The cultivated Passion Flowers, especially *P. incarnata* (bright blue-purple flowers) and others (with red or steel-blue flowers), can be found in almost any inner city area. I have picked them in Morningside Heights (Los Angeles), Isla Vista (Santa Barbara), Escondido, downtown Tucson, Albuquerque, even Las Vegas, Nevada.

So, you are left with two distinct approaches; hiking along streams in javelina and coatimundi tracks in the Guadalupe Mountains or the canyons of the Chisos, or lumbering through the garbage cans and dog doo of your nearest inner-city. Whatever works.

CONSTITUENTS: Harman (passiflorin), harmol, harmalol (roots), harmine, harmaline, 5-hydroxytryptamine, various flavonoids.

COLLECTING: The whole vine, including stems, vines, flowers, and fruit, usually best in early or midsummer. Dry, Method A.

STABILITY: The herb stays relatively strong for at least a year, although it needs protection from any light.

PREPARATION: The tincture, Method B, 1:5, 50% alcohol; a Standard Infusion, 2–4 ounces. Roots (*P. incarnata* and *P. foetida*), Method B tincture, 1:5, 65% alcohol; not very water soluble.

MEDICINAL USES: Passion Flower is a simple, uncomplicated sedative. It slows the pulse, decreases arterial tension, and quiets respiration and pulmonary blood pressure. It is especially well appointed for the sthenic, bull-necked mesomorph, the individual who has an aversion to overt sedation; it has little of the "downer" sensation of other herbs or drugs. It is also especially good medicine for the child or infant with flushed face, peripatetic demeanor, and the stoned crankies with rigid, angry abdomen. It is virtually devoid of side effects in sensible doses and is more calming than narcotic. Some people like an "edge" to their sedatives; Valerian (see *MPOMW*) would probably be their choice (as well as some of our favorite drugs). There is no edge to Passion Flower; it lets you down a couple of notches, and it's then up to you to take it from there.

A cup of the tea for morning sickness usually helps, especially if that morning sickness is at night. The two stinky Passion Flowers, *Passiflora foetida* (native) and *P. incarnata* (urban), have the strongest antispasmodic effects, useful for painful menstruation, stomachache, or diarrhea; their roots are even stronger, but are contraindicated in pregnancy.

The herb tea or tincture is very useful in long-term treatment for functional essential hypertension in the strong, rough-and-tough middle-aged person; one of those that used to work in the field or in a blue collar and is now stuck at a desk as he or she moves up the corporate or government ladder . . . and the blood pressure cuff. See Formula #11 (pg. 139). This has helped folks, from a nurse-midwife in southern New Mexico to a police sergeant in Santa Fe to a book-writing herbalist now living in southern New Mexico.

OTHER USES: The dark-red or purplish fruit is tasty; pick in the early or middle fall.

PERIWINKLE

Vinca major   Apocynaceae

OTHER NAMES: Greater Periwinkle

APPEARANCE: Periwinkle is a large, successful vine with many long stems and widely spaced, oval opposite leaves. These are bright dark-green, smooth, and leathery, an inch or so long. The sterile stems are several feet long, the blossoming stems often shorter, drooping over the longer ones. The lovely flowers are solitary, violet to violet-blue and an inch across.

HABITAT: Periwinkle is strictly a cultivated plant and ground cover, but it does so well in our warm lower altitudes that with a little shade and moisture it is often found growing wild, either escaped on its own or still surviving many years after being planted around a house that has long since disappeared. Usually below 4,000 feet, but sometimes higher in gardens.

CONSTITUENTS: Vincin, vincoside, tannins, and at least sixteen alkaloids.

COLLECTING AND PREPARATIONS: The whole above-ground plant, stems, leaves, and flowers, dried, Method A; the tincture, Method B, 1:5, 50% alcohol, or as a hot infusion, a rounded teaspoon to a tablespoon of the finely chopped herb in water. It is stable when dry.

MEDICINAL USES: This Periwinkle is a strong capillary constrictor, but it also has the additional effect of stimulating peripheral circulation, a rather paradoxical combination. Its most common use is for uterine and rectal bleeding from benign causes; ½ teaspoon of tincture or ½ cup tea every few hours. For heavy menses, especially when the flow is still red after several days or when there is midcycle bleeding from ovulation or uterine fibroids, Periwinkle has few peers, herbal or pharmaceutical. It combines well with Cotton Root Bark, Shepherd's Purse, or Trillium fresh root tincture, 20 drops (wait until *Medicinal Plants of the Pacific Coast*).

In case of a nosebleed from chemical irritations or allergies, the tea, taken 1 cup every three or four hours, will staunch the flow effectively. It has less value for nosebleeds from trauma, hypertensive episodes, or manic nose picking.

Like Clematis, Periwinkle is unpredictable in cases of migraine and other severe headaches; nonetheless it sometimes is *the* answer, where little else aids. The best specific indications I know of are sharp, left- or right-sided headaches, adrenaline-type stress, with pounding in temples upon standing suddenly, sympathomimetic drug excesses (such as cocaine, amphetamines, or even coffee) and bloodshot eyes with a sweaty forehead. Moderate amounts, as recommended above, are mildly hypotensive, and some of its systolic-lowering effects may account for these benefits.

85

CONTRAINDICATIONS: Other prescription medicines (it has a complex pharmacology and may cause drug inter-reactions), pregnancy (to be on the safe side), and chronic constipation (it may decrease gastrointestinal secretions).

PINEAPPLE WEED

Matricaria matricarioides   Compositae

OTHER NAMES: *M. suaveolens, M. discoidea,* Wild Chamomile, Manzanilla del Barrio

APPEARANCE AND HABITAT: Pineapple Weed is a close relative of regular old Chamomile, and many city dwellers consider this the same plant, especially since their closest contact with true Chamomile is the jar in the health food store or co-op. The commercial herb looks like it; the white daisy petals of the official plant turn beige and nondescript on drying and fall off the flower heads, leaving the obvious, little yellow cone-centers. Pineapple Weed has no ray flowers but is similar in use to the European herb, variously called German Chamomile, Hungarian Chamomile, or (if you live in Germany) English Chamomile. If you live in England, Roman Chamomile is a close relative, *Anthemis nobilis*; if you live in Italy, Roman Chamomile is called German Chamomile, sometimes just True Chamomile . . . etc.

Ours is just Pineapple Weed, probably because it smells a little like Pineapple, is a common city weed, and you can call it Wild Chamomile if you want, it's fine with me. If the Europeans, who *invented* Chamomile, can't decide whose herb they drink and botanists can't decide which species or genus they are talking about, we still have our little vacant lot flower. It is most common in California, but grows abundantly in such diverse places as Tucson, Albuquerque, El Paso, and Denver, blooming from March to June, but as early as February in San Diego or Austin. It has little pinnate leaves and from one to a dozen conical, yellow flowers. A durable weed, it may grow two inches high where school kids trample it down, or a full, luxuriant foot high in a moist, shady place by a chain link fence or in the rich soil of racing horse stalls.

COLLECTING: The whole annual, rootlet and all, while the oldest central flower head is still yellow; dry, Method A.

PREPARATION: A Standard Infusion or just brew it any old way. Externally, the plant steeped in olive oil, one part finely chopped dry herb in three parts (by volume) of oil; steep for a week, squeeze out.

MEDICINAL USES: Italian studies with both *Matricaria camomilla* and *M. matricarioides* have shown that the tea and the steeped oil (Oleum Chamomillae Infusum is the technical term) are mild but distinct anti-inflammatory agents when applied topically. Emollients such as the steeped oil soften and relax the tissues, diminish tension and pressure on the sensory nerves, dilate the capillaries (both blood and lymph), and generally protect the inflamed surfaces against further irritation. Taut, cranky, red-faced, belly-sore children respond

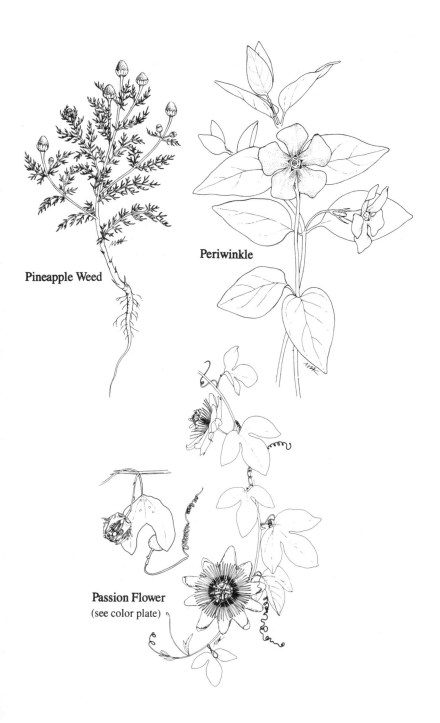

Pineapple Weed

Periwinkle

Passion Flower
(see color plate)

especially well to the tea or the oil rubbed all over them . . . as well as to parental attention.

The tea is mildly relaxing, with just enough aromatic constituents to anesthetize the mouth and stomach lining and to be antispasmodic for stomachaches, gas pain, and teething. As a sedative it mixes well with California Poppy, Passion Flower, Wild Oats, Prickly Poppy, or Skullcap. (The last two are described in *MPOMW*.) As an intestinal or uterine antispasmodic, mix with Desert Lavender, Chimaja, or True Sage.

PIÑON

Pinus spp.   Pinaceae

APPEARANCE AND HABITAT: Two characteristics separate the Pines from other conifers such as spruce, fir, and Douglas fir. In Pine trees, the needles are in bundles of two more with a tiny papery sheath surrounding the base of the bundles. The single exception is *P. monophylla*, a variant of the typical common Piñon of the Juniper/Piñon belts. Secondly, Pine pinecones are woody and stiffer than those of spruces, which are papery scaled and very flexible. The cones of firs stand upright in contrast to the hanging positions of the other conifer cones, and Douglas fir cones contain little three-pointed membranous bracts between the cone scales. The various Pines are found from 5,000 feet (or lower) to, and often forming, the timberline. The little Piñon trees are the lowest in elevation, needing the least moisture, followed by the Yellow Pine and Ponderosa belts, then a whole gaggle of various Pines in the main forest areas.

COLLECTING: The inner bark, Method B, the needles collected in single branches and hung over a newspaper until dry. If the pitch is needed in any quantity, it should be melted at as low a temperature as possible and poured through a metal sieve warmed over a burner to separate the good from the bad. The sieve may be cleaned with turpentine or may be totally unusable, depending on your tenacity. For less fussy types, the pitch may be saved as collected, adhering bark, dirt, bugs, and all.

MEDICINAL USES: Pine needles make a very pleasant tea simply for the taste and have a mild diuretic and expectorant function as well. The inner bark boiled slowly for tea and sweetened with honey is still stronger as an expectorant, useful after the feverish, infectious stage of a chest cold has passed. The pitch is the most specific of all; a piece the size of a currant is chewed and swallowed. This is followed shortly afterwards by strong, fruitful expectoration and a general softening of the bronchial mucus. This remedy is especially useful for children. The pitch also has some value as a lower urinary tract disinfectant but would be inappropriate for use when kidney inflammation is present. In New Mexico, *trementina* (pitch) is warmed slightly over a stove or campfire and applied to splinters, glass, and other skin invaders, allowed to set, and peeled off, carrying the problem with it. While gathering near Questa one day, I encountered an elderly patriarch named Joe Rael (it was mid-afternoon and he was working on his third cord of wood for the winter) who had run a splinter half-way up his arm . . . or so it seemed. Cursing in obscure Spanish, he grabbed some pitch, warmed it over a cigarette lighter in a crushed beer can, and slapped it on the wound, waited a moment, and plucked out the splinter with the pitch. I tried the same thing the next week, and got a blister for my troubles. An acquired technique, I guess. Sometimes it takes several days to facilitate the gradual isolation and extraction of the splinter, so stick with the pitch during several applications.

PRICKLY PEAR

Opuntia phaeacantha, O. compressa, O. polycantha, etc.   Cactaceae

OTHER NAMES: Beavertail Cactus, Nopal, Tuna

APPEARANCE: Prickly Pear and Beavertail Cacti, by whatever their Latin names (and these change with the seasons), look alike and are easily identified. They have no leaves (except at the start of new growth), they have spiny, thickened stems that form the body of the plant and produce lovely yellow, orange, and red, rose-like flowers in the spring. These mature into the prickly pears (nopalitos), also yellow, orange, red, or purple. Chollas are sometimes put into the same genus (*Opuntia*) as Prickly Pears but have rather different sap constituents and should not be considered a substitute. Cholla flowers, however, are interchangeable with those of Prickly Pear and can be gathered with them for the same uses.

HABITAT: Everywhere and anywhere, from British Columbia to the Appalachian Mountains; especially common, of course, in the Southwest. A family of plants totally native to the Americas, except for a little, mistletoe-like cactus found in areas around the Indian Ocean (spread there by birds), the Cacti presumably evolved from the Rosaceae. Although even the most common of the Prickly Pears are protected plants in Arizona, the reality is such that those listed above, and others, are widespread, ubiquitous, and could care less if you purloined a few flowers and several pads. They survive under a ferocious variety of climates and regenerate from pads, root callouses, and seeds, so there is really no moral question about some judicious pruning from healthy plants.

COLLECTING: There are several spineless varieties, both wild and cultivated, but even the normal spiny/grumpy pad can be skinned from the soft inner flesh with a fillet knife (and a little care); the spineless pads can also be purchased in the produce sections of many of our markets. It is a traditional food in Mexico and parts of the Southwest.

The flowers of most species are surrounded by glochids, those little spine-hairs that get into your skin and are so hard to remove. These are not spines (which are modified branches) but actually leaf hairs, with little retorsed barbs and a brittle, easily broken structure. Harvest the flowers with care. As a rule, the more reddish the flowers the stronger the constituents present. Dry, Method B.

PREPARATION: The juice can be prepared by kibbling the skinned inner flesh, pureeing in a blender, and drinking it as is or squeezing through a cloth by hand, clothes wringer, or hydraulic press. Both the slurry and the expressed juice work the same way, but some people object to the unpleasant mucoid consistency of the slurry. The juice can be preserved with 25% grain alcohol added, but this should be for the urinary tract uses; that much alcohol can be a problem for those with hyperglycemic conditions. Raw fillets of the flesh are best for poultice use. The slurry, juice, and fillets can survive refrigerated for up to a week. The flowers are best as a simple infusion, well strained before drinking to prevent swallowing those glochid hairs.

MEDICINAL USES: The filleted pads are beautifully effective drawing poultices. Place them against the parts injured, cover in gauze, and tape to the skin; remove after several hours. Contusions, bruises, and burns that are engorged and tender contain much feral blood and disorganized interstitial fluid. The mucopolysaccharide gel in the Prickly Pear flesh is strongly hydrophilic and hypertonic; some of the fluid exudates that build up in the injury are absorbed osmotically through the skin and into the cactus, while the gel softens the skin, decreases the tension against the injury, and lessens the pain. This is, by the way, how Aloe Vera works. Finally, small, skinned sections may be held between the gum and cheek to lessen the pain and inflammation from gum infections and mouth sores.

In Mexico, the juice is widely used as an anti-inflammatory diuretic. When there is pain on urination, with a continuing dull ache in the urethra and bladder well after completion, the juice can be taken to decrease the pain. Use teaspoon doses every two hours until the pain is gone. This, of course, does not affect any bacteria that may be causing the pain; it is taken to relieve the inflammation. Even if a prescription antibiotic is needed, the juice will still help to lessen the pain until the medication takes effect.

For Honeymoon Cystitis, when two frenzied folks have partaken of each other in such giddy abandon that they both waddle crablike and bowlegged for days, the juice helps sooth irritated urethral tissue; take a teaspoon or two every two hours until both can walk at full height for fifty feet.

Recent studies with adult-onset diabetics in Mexico have shown that the folk use of Prickly Pear for diabetes is clinically verifiable. There was a clear hypoglycemic effect on the obese, insulin-resistant patients, with simultaneously lowered serum levels of low-density cholesterols and triglycerides. This is a rather unique effect, since I have taken many hypoglycemic herbs, along with friends and students ("provings," they used to call it), and Matarique, Prodigiosa, and Tronadora all lowered the blood sugar in healthy persons; Prickly Pear did not, and seems to be hyperglycemic-specific. The effective doses used in the clinical study from Mexico averaged 4 ounces of the juice per day. This will entail a moderate amount of effort on the part of the diabetic to squeeze this much, regularly. Conversely, adult-onset diabetes is a chronic disease and not a self-limiting one; any regimen must be extended. With the accuracy of the currently available self-diagnostic tests for serum glucose levels, it would certainly seem to be a useful alternative to try, especially when you consider the side effects encountered in oral hypoglycemic medications. As

Añil del Muerto

Acacia

Agave

Anemone

Cañaigre

Cenizo

Buckwheat Bush

Condalia

Desert Senna

Chaparro Amargosa

Desert Willow

Chaparral

Desert Lavender

Elephant Tree

Hollyhock

Jojoba

Milk Thistle

Night Blooming Cereus

Puncture Vine

Ocotillo

Prickly Pear

Passion Flower

Ratany

Ratany

Sangre de Drago

Sangre de Drago

Sangre de Drago

Tronadora

Sumach

Yerba Mansa

mentioned before, the fresh juice is the preparation needed (the slurry, if you don't mind the consistency); the alcohol-preserved juice in such quantities as the fresh plant is used in would supply enough alcohol to create a blood-sugar problem.

The dried flowers, high in flavonoids, are a useful and elegant treatment for capillary fragility, particularly when the mucosa have shown stress and inflammation for any period of time. Examples of conditions where it is helpful would be chronic colitis, pulmonary problems like asthma and mild brochiectasis, benign prostatic hypertrophy, chronic vaginitis, and diverticulosis. It won't reverse the condition, but it will help to strengthen the capillary beds and the submucosa, enabling the tissues to regenerate better. For starters, try one whole dried flower in a well-strained infusion, three times a day.

OTHER USES: The ripe fruit are delicious! The glochid hairs are not! Skewer the fruit on a long knife or hardened stick and roll it back and forth over an open fire, gas flame, or propane torch until they are *all* singed off and enjoy the sweet, tangy (and slightly poached) flavor. Even the grossly seedy varieties taste good; squeeze their juice through a cloth and make jam with it. Cook them up in a little water and honey, strain out the seeds, and you have a tasty pancake syrup. High in ascorbic acid and bioflavonoids, don't heat too high lest these nutrients be lost.

PRICKLY POPPY

Argemone spp.   Papaveraceae

OTHER NAMES: Thistle Poppy, Cardo Santo, Chicalote, Mexican Poppy

APPEARANCE: A peculiar plant, Prickly Poppy has all the superficial resemblances of a thistle; prickly leaves and stems, blue-green, inconspicuous foliage, and a mature height of one and one-half to three feet. The flowers, however, are big, showy poppies with five, white, papery petals and a golden yellow center and they bloom from late spring to autumn. Totally unexpected if you had been observing the plants before flowering: scruffy and weedy, they seem the sort one instinctively avoids. The stem-clasping, wavy leaves are from three to eight inches long, and the stem is stout and well armored. All parts of the plant have a yellowish, milky sap that turns black in contact with the air. The flower buds are three-horned and well barbed; the flowers themselves lose their petals one at a time, to be eventually replaced by ridged pods with a tip of black sap. The blooms are in clusters, the center being the oldest and, when fruiting, the earliest to open. The football-shaped pods contain poppy seeds that are milky white at first, becoming mottled brown-black when mature; the seeds are gradually sown by the wind as the sides of the pod capsules curl back. Annual, sometimes perennial.

HABITAT: From sea level to 8,000 feet, Prickly Poppy is found throughout our area. It has no predictable habitat, although it is common in disturbed earth, overgrazed pastures, and along roadsides in the middle deserts.

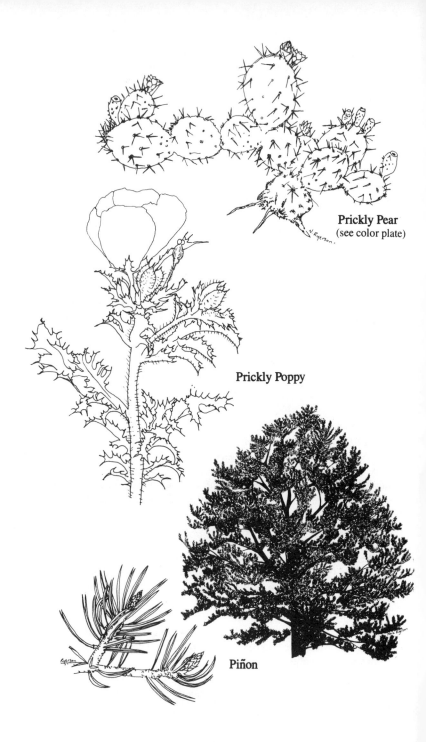

Prickly Pear
(see color plate)

Prickly Poppy

Piñon

CONSTITUENTS: Protopine, berberine, artarine, allocryptine, argemonine, sanguinarine, and other closely related alkaloids.

COLLECTING: Method A: the seeds when brown and before the pod has opened, the plant when in flower. Since the sap is narcotic enough that a few moments of handling the plant results in skin insensitivity to the spines, you may wish to tough it out and gather it naked-skinned; otherwise use gloves for this nasty . . . thick gloves. There is probably some easy way to winnow the seeds, but I have not found any, except drying the pods in the sun, covering them with burlap, and stomping them to death. The seeds can then be sifted. For tea, the leaves should be stripped when dry (thick gloves) and chopped up for storage.

PREPARATION: For a burn and scrape salve, grind the seeds down into an oily paste, warm over low heat in twice the volume of petroleum jelly, hydrous lanoline, or lard, remove from heat, repeat the melting a few times for a day or two, melt once more, and strain into jars. The herb for tea, standard infusion, 2–4 ounces as needed.

MEDICINAL USES: The juice of the fresh plant has a rubifacient and somewhat caustic effect and was used for burning off warts; it works sometimes. The tea is analgesic topically and can be applied freely to sunburns and abraded tissue to relieve both the pain and the swelling. Used internally it has a feeble but distinct opiate-like effect, and the tea can be used for urinary pain and mild prostatitis as well as simple sedation. It combines well with Skullcap for neuralgia, sciatica, or other nerve pain; too much tea and you wake up with a groggy hangover . . . the kind you get from going to a bargain basement opium den where they cut the latex with instant coffee and ground bakelite. The tea also has distinct antispasmodic effects and, alone or with Silk Tassel, can really modify the distress of diarrhea cramps. It also helps palpitations from indigestion, haitus hernia, caffeine excess, and hormone wobbles caused by premenstrual conditions or menopause.

The ointment made from the seeds is an excellent soothing and healing cover for severely sunburned skin and large-area, low-severity burns from hot water and the like. As mentioned, use petroleum jelly, lanolin, or rendered lard, not vegetable oil and beeswax; it is important that the seed oils not become rancid, which will occur in the vegetable oil. The seeds are a strong castor-oil-like cathartic, a teaspoon or two crushed in water and drunk. They have a sedative effect when eaten and have traditionally been smoked alone or with tobacco, but crush them a little first . . . they can pop in your face when lit.

OTHER USES: They make a fine food, lightly roasted on the stove or in the oven and used in the same way as poppy seeds, dusted on bread, mixed in granola, etc.

PRODIGIOSA

Brickellia grandiflora and
others   Compositae
OTHER NAMES: Hamula, Atanasia
Amarga, Bricklebush

APPEARANCE: *B. grandiflora* is the most attractive and most widely used of the genus (at least as a medicine). It is a large, many-branched herb-bush, up to three feet tall. The leaves are dark green and purplish, paired except up toward the flowers, where they start to become alternate; they are from two to three inches long, triangular, and moderately toothed. The plant can easily resemble Catnip or some large-leaved Sage, but its flowers are totally different, forming clusters of drooping, petal-less composite flowers hanging from the ends of the stems and usually partway down one side. These flowers are cream-white to yellow, rather nondescript, and will number dozens on a fully flowering stem. Not a raging beauty of a plant, compared to other *Brickellias* this is the jewel, what with its dark green, silvery-purple foliage, large leaves, nodding flowers, and its vague sense of upright, articulate usefulness.

HABITAT: Prodigiosa grows in the lower foothills of the taller and warmer ranges, from eastern Sierra Nevada down through central and southern Arizona, east and north through the lower canyons of the Jemez, Sangre de Cristo, and southern Rocky mountains, Guadalupe Mountains in New Mexico and Davis Mountains in west Texas, and east to Arkansas. Other species, such as *Brickellia incana* (silvery foliage, conical maroon flowers, Mojave and Colorado desert washes) and *B. californica* (more rounded leaves that are three-ribbed and rough textured, flowers off-white and nodding, sweet scented and drab), can be used as well.

CONSTITUENTS: Various labdanes, a flavoneglucoside, penduline and penduletine, a dehydronerolidol derivative, a benzofuran, and miscellaneous diterpenes.

COLLECTING: The flowering branches, bundled, dried.

STABILITY: Good for at least a year; too much longer and it loses in strength.

PREPARATION: The tea, Standard Infusion, 2–4 oz. doses; the tincture, Method B, 1:5, 50% alcohol, ¼ – ½ teaspoon doses.

MEDICINAL USES: Prodigiosa has three distinct uses: 1) lowering blood sugar in certain types of diabetes; 2) stimulating hydrochloric acid secretions by the stomach; and 3) stimulating bile synthesis and gallbladder evacuation.

Hyperglycemia: It mildly inhibits epinephrine's stimulation of the liver to produce glucose. Epinephrine (adrenalin) stimulates the liver to break down stored glycogen directly into blood sugar. This is fine if you are running away from a saber-toothed tiger, but not so fine if you are sitting at your desk grumbling about your job. It always seems easier to release stuff into the blood from storage than to send it back where it came from if it isn't needed. With stress elevation of glucose, insulin is secreted to get the levels down, as the brain is unable to control its combustion and goes weird on you if it has too much fuel. The excess sugar is shunted into other cells of the body for them to store, either as fat or starch. A few years of this and those accommodating cells that have been taking the load off the brain (the cause of the problem) are so engorged with fuel that their metabolism is impaired. Pretty soon *they* start refusing the insulin commands, and the blood sugars start to stay elevated. This is called insulin-resistant diabetes and is the most common type of adult-onset diabetes. Since endocrinologists and diabetic specialists get into fistfights at conventions over how, why, and when diabetes even happens, I don't feel too ashamed in saying that I have no idea of the mechanism involved, but a cup of Prodigiosa tea in the morning and a cup in the afternoon is a regimen followed by thousands of Native Americans, Mexicans, and New Mexicans to help control their condition. In the beginning it may be helpful to use some Matarique along with the Prodigiosa, as the latter is stronger and more vigorous. The Prodigiosa is the better daily herb.

Stomach: Prodigiosa has clear and strong effects on the stomach lining, increasing both the quantity and acidic quality of secretions. In the condition of achlorhydria, when there is poor manufacturing of hydrochloric acid, food, especially protein and butter fats, must stay far longer in the stomach before being let into the duodenum. The longer the food stays in the stomach, the more likely acid indigestion will happen, and the more likely you are to become hypersensitive to proteins in your food; heavy drinkers often have impaired stomach function. And, the older we get, the more likely we are to have this problem, no matter what our recreational habits are. The best way to use Prodigiosa is a mild cup of the tea or half teaspoon of the tincture in water in the late afternoon; the stomach is usually the most inhibited in the evening, and this helps us better digest dinner . . . which is usually our heaviest meal.

Gallbladder: Long credited with "flushing" gallstones out (highly unlikely), Prodigiosa has beneficial effects on fat digestion in general and gallbladder evacuation specifically. This helps prevent gallstones or a gallbladder attack, since the longer bile is retained in the gallbladder, the more likely it is to precipitate its cholesterol into stones.

OTHER USES: In Mexico the strong tea is used like Escoba de la Vibora, as a strong bath for acute arthritis. It has also been shown to inhibit the enzyme called lens aldose reductase, thus potentially preventing or helping cataracts. Unfortunately (for me, at least), the monographs in journals of experimental pharmacy and natural products chemistry and pharmacognosy are almost completely lacking in information on applications, so I haven't the foggiest idea how that could be of help to potential cataract sufferers . . . a tea? an eyewash? injection? Oh well.

CONTRAINDICATIONS: Acute cholelithiasis (stone blockage), hyperchlorhydria (too much stomach secretion), and insulin-dependent diabetes.

PUNCTURE VINE

Tribulus terrestris   Zygophyllaceae

OTHER NAMES: Terror of the Earth, Goats' Heads, Little Caltrop

APPEARANCE: A terrible, disgusting, gross, little ground-covering weed, with long, vinelike pinnate leaves, pleasant yellow flowers, and terrible, disgusting, gross little four-sided seed pods that break apart into three-sided capsules that maim pets and crawly children, puncture bicycle tires (hence one of the names), and, if you step barefooted on one, give rise to yet another common name, "&#&%!."

HABITAT: Dry vacant lots, gravelly junk-soil, all kinds of southwestern city, suburban, and rural waste-places. I say waste-places because too many Puncture Vines can cause anyplace to become a waste-place quickly. Yucky.

CONSTITUENTS: Diosgenin, ticogenin, hecogenin, the flavonoid astragalin, harman (leaf), harmine (seed).

COLLECTING AND PREPARATION: The whole plant, preferably while the seeds are still green, dried any damn way you feel like (excluding flamethrower). While you are gathering the plant (not a major aesthetic experience otherwise), remember that every seed you take up is one less for bicycle tires, kids, and pets, one less seed to grow next year, one small step for mammalkind. Just be sure to bottle or bag the dried plant so that the caltrops don't get around the house. The dried plant is rather stable; so powder it up as best you can. The woody shell of the seeds is poorly soluble in water . . . or anything, for that matter. Isn't it nice to know that this (expletive) has some use?

MEDICINAL USES: Some rather extensive studies have shown that the seeds and, to a lesser degree, the foliage is a useful early treatment for elevated blood fats, including cholesterols. It helps prevent or lessen the severity of arteriosclerosis and atherosclerosis. The reasonable dose is ½–1 teaspoon of the powdered plant in hot water for tea, morning and early evening. Puncture Vine is also useful in mild essential hypertension; it has a negative chronotropic effect on the heart, slowing the adrenergic stimulation, acting as a catecholamine liberator, and increasing the force of myocardial contractions. This means a slower, stronger, more well-defined heart function, with greater relaxation between contractions and a lowering of the diastolic pressure. It also acts as a sensible diuretic, moderately increasing sodium loss and decreasing overall interstitial and abdominal fluid retention.

CONTRAINDICATIONS: Kidney disease, more serious cardiovascular disorders, and mild or severe liver disease. Keep the doses small and reasonable; large doses can irritate the kidneys and cause dizziness. If the small doses don't help, it isn't the right therapeutic for you; more is not going to help, only hinder. Even in powder, Puncture Vine is not benign.

RATANY

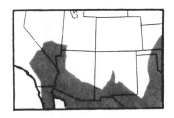

Krameria grayii, K. parviflora, K. lanceolata   Krameriaceae
(Leguminosae)

OTHER NAMES: *K. secundifolia*, Rhatany, White Ratany, Crimson Beak, Chacate, Crameria

APPEARANCE: It's easy to identify the Ratanies when they are in bloom. They have iridescent, little-dime-sized magenta flowers with three regular petals and two sort of fused together; they are a little orchid-like, a little sweet-pea-like with a bit of Mint family and *Lobelia* thrown in. The leaves of our species are narrow, entire, and opposite; the seeds form little barbed burrs. These are considered semiparasitic plants on the roots of bushes. No Ratany gets very big; *Krameria grayii*, fuzzy and whitish-leaved, may get two feet high, a pleasant mass of crisscrossing semispined branches when not blooming; *K. lanceolata* may get two feet long (it's a drooping semivine), but that's it. Mostly they bloom from April to June, although the White Ratany (*K. grayii*), with its extreme low-desert habitat, blooms whenever there is moisture, even in January, although May brings out the most fragrant blooms. Although sometimes classed with the Pea family, recent studies have shown it to be related to the Milkworts (Polygalacea).

HABITAT: Sporadic in all areas below 5,000 feet; usually in open, lovely desert grasslands, with *K. grayii* liking the lowest desert and *K. lanceolata* the higher altitudes of what is left of the southwestern prairies. There are very few Ratanies in the Great Basin, Utah, or Colorado, but elsewhere in our area it is a well-established family of oddities.

CONSTITUENTS: Krameriatannic acid, krameria-red (a phlobaphene), rhatanine, N-methyl tryosine.

COLLECTING: The root is strongest, the foliage, weaker but with the same effects. Our species have smaller roots than the former official Krameria U.S.P., but, with stronger and thicker bark, was once considered a better source than the official species, particularly *K. lanceolata* (*K. secundifolia*). It is fortunate that it was never the official Krameria; otherwise, our three little species would have become as endangered as some of our other drug plants like Pinkroot, Virginia Snakeroot, and American Ginseng. In the first half of the twentieth century, *tons* of Ratany roots were harvested every year in Peru, Bolivia, and Brazil. Ours could not have survived gathering that intense.

PREPARATION: The dried root, tinctured, Method B, 50% alcohol, adding 5% glycerin when finished. The foliage is reserved for tea, brewed as a Strong Decoction.

MEDICINAL USES: Ratany is one of our best plant astringents and topical hemostatics. For sore gums, abscesses, and mouth sores, use alone (tincture or tea) or combine with the tinctures of Elephant Tree, Yerba Mansa, Sumac, or Myrrh. See Formula #12 (page 139).

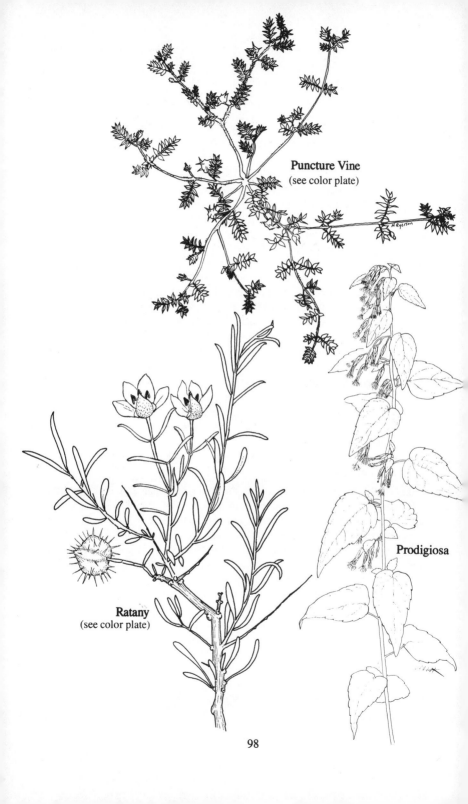

Puncture Vine
(see color plate)

Ratany
(see color plate)

Prodigiosa

98

Gargle the tea or diluted tincture for acute or lingering sore throat; it can be combined for this purpose with Yerba Mansa or Echinacea. For diarrhea, combine with Silk Tassel (for cramps) and Echinacea (immunostimulant), and with either Trumpet Creeper, Desert Willow, or Tronadora (for Candida) and Chaparro Amargosa (protozoas). See Formula #13 (pg. 139). For a hemorrhoidal salve and rectal fissure ointment, use either alone or with Echinacea flowers; as a salve, Method A; this also helps vaginal sores.

Ratany will decrease internal bleeding, but it can cause constipation and in fact is not usually as effective as Shepherd's Purse or Canadian Fleabane. These have the advantages of being not constipating and more common. Generally, Ratany is used best for sores and ulcers of the beginning (mouth, throat) and the end (colon, rectum) of the gastrointestinal tract.

RED ROOT

Ceanothus spp.   Rhamnaceae

OTHER NAMES: Deerbrush, Buckbrush, Tobaccobrush, Mountain Lilac, Snowbrush, California Lilac, (New) Jersey Tea, Palo Colorado, Chaquerilla

APPEARANCE: Although a widely varied and interbreeding genus, several characteristics are nearly universal. The small seed pods resemble a horned acorn (except *C. greggii* and a few relatives that are more smooth capsuled), distinctly three lobed and tending to be shaded darker where they face the sun, as if spray painted from that side. The small, delicate flowers are borne on showy little puffs at the ends of straight stems that tend to stand out from the branches at sharp, almost right angles (except *C. integerrimus,* which has a languid, drooping panicle). In California, particularly along the coast, there are some attractive species that grow as smallish trees with beautiful lilac, pink, and purple flowers; the more common types have white or cream-colored flowers, small dark green or olive-green leaves, and sparsely leafed or leafless branchlets that double as blunt "thorns." *C. cuneatus, C. greggii,* and *C. fendleri* are the most common species of this type that are found in the middle desert up through the dry Ponderosa and Yellow Pine canyons; the first two are small shrubs from two to six feet high; *C. greggii* (Desert Ceanothus) has lighter bark and somewhat inflated, stubby branches, flatly toothed leaves, with *C. cuneatus* having more rounded leaves and more downy hair, and the two interbreeding as they meet. *C. fendleri* is distinctly blue-green leafed, covered in vague, brittle spines and growing as a scraggly ground cover to a small shrublet, up to three feet high. The typical Red Root flowers are lacy, fragrant clusters that form a soapy foam if rubbed in water. The roots of *C. greggii* and *C. cuneatus* give off a slight wintergreen scent.

HABITAT: The lower-growing group (*C. greggii* and *C. cuneatus*) are found on the dry slopes of the desert-mountain interfaces, along with Manzanita, Condalia, and Oak, usually in the shrub-stubble of the lower slopes, from dry coastal and inland mountains all the way through the Mojave, along the middle elevation of the Mogollon Rim, and from the southern half of Arizona, New Mexico, West Texas, and northern Sonora and Chihuahua states, from 3,000 to 6,500 feet. *C. fendleri* and *C. integerrimus* and the California group fill in all the rest of the moister and midmountain areas, which is why I didn't bother with a distribution map; in the West there is a Red Root almost anywhere except the lowest desert.

CONSTITUENTS: Leaves: Nonacosane, 1-hexacosanol, velutin, traces of caffeine reported in several species; Root and Bark: Betulinic acid, ceanothic acid, ceanothenic acid, methyl-salicylate (several species), and the alkaloids ceanothine, ceanothamine, integerressine, integerrenine, integerrine, and americine.

COLLECTING: Gather roots in the late fall when their color is darkest or in the very early spring before flowering; at any time, however, they will have some value. The plants are tough and wiry, the roots doubly so. The two desert species have denser and more organized single-root masses and are best dug after recent rain, with moist gravel and dirt around them; the higher and moister species tend to have more dispersed, runner-like roots, branching underground from a ferocious taproot that would feed a twenty-foot oak. All the Red Root roots entail sharp, stout clippers or branch shears to split them apart while fresh; after drying, use a diamond-tipped bandsaw or jackhammer.

STABILITY: The leaves are stable for up to a year; the root and root-bark of the two desert species are moderately stable, although if they are tinctured dry, it should be within about two or three months of gathering. The higher desert species and coastal species are stable for well over a year for tincturing or tea.

PREPARATION: Fresh root tincture, Method A, especially *C. greggii* and *C. cuneatus*. The root may be tinctured quite well in the dry state, 1:5, 50% alcohol, *except* grinding it down from dried chips to a coarse powder is a major difficulty . . . which is why I have come to prefer the fresh tincture. The dry root is best made for tea as a cold standard infusion. Dose of either tincture, ½–1½ teaspoons; of the infusion, 2–4 ounces.

MEDICINAL USES: Red Root has a slight but distinct tendency to stimulate the quality of electrical repelling that occurs between the circulating blood proteins and the facing wall of red blood vessels. Forcing the proteins and the protein walls of red blood cells to the center of the capillaries allows a better surface interchange between the inside surface (toward the blood) and outside surface (toward the interstitial fluids) of the endothelial cells that form the blood capillaries. The old medical axiom still holds true: the blood feeds the lymph (interstitial fluid) and the lymph feeds the cells and tissues of the body. Since Red Root helps increase the efficiency of transport of nutrients from the blood across the capillary cells to the lymph, the increased charge potential is manifested as increased efficiency of lymph transport of waste products, away from the cells and eventually back to the blood and the liver . . . the reversing of the feeding process. This makes Red Root a lymphatic remedy, stimulating lymph and interstitial-fluid circulation.

Red Root's many applications include preventing the buildup of congested fluids in lymphatic tissue as well as clearing out isolated fluid cysts (internal blisters) that may form in some soft tissues. It will help reabsorption of some ovarian cysts and testicular hydroceles when combined with Dong Quai (cured Chinese Angelica Root) or Blue Cohosh and Helonias Roots (commonly available from herb or health food stores). In Oriental medicine these conditions are deficient (cold), and the accompanying herbs increase either the rate of ovary- or testes-metabolism (Dong Quai) or the rate of arterial and veinous circulation in the pelvis (Blue Cohosh and Helonias); they are thus stimulating (hot). The Red Root helps to increase the fluid removal from the cyst lymphatically. For breast cysts that enlarge and shrink with the estrous cycle and have been diagnosed medically as such, combine the Red Root with Cotton Root, Inmortal (see *MPOMW*), or 3–5 drop doses of Phytolacca tincture (fresh Poke Root extract available from an N.D. or, as a mother tincture, from an M.D. homeopath).

In a more prosaic vein, Red Root is an excellent treatment for tonsil inflammations, sore throats, enlarged lymph nodes, and chronic adenoid enlargements. For adults, two tablespoons of the root can be boiled for twenty minutes in a quart of water and refrigerated, a third of the quart drunk an hour before each meal (as well as the cold infusion mentioned above, which, although a bit better an extraction, is a longer process). Further, a few days of this regimen will generally reduce the size of an enlarged spleen.

For those who use Oriental diagnostics, a rounded tablespoon of Red Root and a scant tablespoon of Vervain brewed similarly and taken for several days will help clear the meridians of the torso, pelvis, and legs. If these are long-standing blockages and the meridians are either over- or undersensitive, this treatment helps to clarify both diagnosis and therapy. This is an empirical observation on my part, and I haven't the foggiest idea why it helps.

Red Root is an excellent home remedy for menstrual hemorrhage, nosebleeds, bleeding piles, hemorrhoids, and old ulcers, as well as capillary ruptures from vomiting or coughing. The tincture of California Lilac was used by the homeopath Boericke around the turn of the century for sore throat, inflamed tonsils, sinus inflammations, and diphtheria, both internally and as a gargle. The leaves make a pleasant beverage for drinking (brewed lightly) and, stronger, it can be used as a hair tonic.

SAGE

Salvia spp.   Labiatae

OTHER NAMES: Black Sage, Red Sage, Scarlet Sage, Crimson Sage, Purple Sage, White Sage . . . etc., etc.

APPEARANCE: In the Southwest there will always be some Sage nearby. They are medium to large herbaceous bushes with opposite, usually wrinkled or rugose leaves varying anywhere from smooth, silvery white to wrinkled gray-green to dark shiny-sticky green. The flowers form either bracted spikes or the distinct cyme-balls, with the flower colors ranging from pink to red, blue, purple, and even white. The varieties of Sage are so many in our area that you may need a local flower guide to help you further, or at least the nearest arboretum or botanical garden. One rule of thumb with this varied family . . . if it is at least moderately smelly, it's a useful remedy.

Many people confuse true Sage with some of the Wormwood family (*Artemisia* genus), such as Sagebrush, Silver Sage, and others. These Wormwood "Sages" with their often gray color and sagelike smell result in much confusion; they do not have opposite leaves, their flowers look nothing like the flowers of the genus *Salvia*, and they are intensely bitter. If you use one of the Wormwoods for Sage dressing, you are guaranteed to ruin even the biggest, oldest, gamiest tom turkey you can catch, let alone the turkoid fowls of the supermarket; some of the native Sages might taste a bit peculiar in stuffing, but they would still be recognizable relatives of Garden Sage.

HABITAT: From the surf to the Mojave Desert, the native Sages *own* the desert hillsides and lower mountains of southern California, often forming nearly pure stands of several mixed species. Here we have *Salvia clevelandii, S. mellifera* (Black Sage), *S. leucophylla* (Purple Sage), and the strongest medicine (extending almost to Arizona,) *S. apiana,* the well-known White Sage. *Salvia dorrii,* with its blue flowers, silver stems, and blue-green leaves, can be found sporadically but abundantly where located in sandy washes and alluvial fans from southeastern Washington all across the lower elevations of the Great Basin, south into the eastern Mojave, east nearly to Flagstaff. *Salvia arizonica* (brilliant dark blue) and *S. lemmoni* (brilliant red) are two colorful and gently scented little perennials found in the upper and middle slopes of the desert mountains from southern Arizona, east to Texas, south to the Sonora and Chihuahua states. *Salvia columbariae* (Chia), *S. subincisa,* and *S. reflexa* (Chan) are the ephemeral annuals of the genus, appearing along the roadsides (the first one below 4,000 feet only, the others up to 7,000 feet).

CONSTITUENTS: Isopimaric acid and related diterpenes, histamine, and a wildly assorted group of aromatics, depending on the species, growing conditions, and time of year.

COLLECTING: Bundled and dried, Method A; best in flower.

STABILITY: As long as the dried herb has its characteristic scent.

PREPARATION: Tinctured, Method B, 1:5, 50% alcohol, standard infusion, cold infusion.

MEDICINAL USES: Sage decreases excess skin secretions when taken internally, acts as a reliable astringent for gargling and washing externally, and has clear antimicrobial effects. The classic traditional use of Sage for helping to wean infants from breastfeeding works fine if the tea is drunk cool and the breasts are washed with tea as well. Several goatkeepers I know of use the tea whenever a particular female needs her milk slowed or stopped. They have found the plants I brought them from Arizona and New Mexico (*Salvia dorrii, S. lemmoni, S. arizonica,* and *S. pinguifolia*) to be better than the various California species, with the exception of *S. clevelandii.* Conversely, the various Sages work well, drunk hot, to stimulate sweating and a more comfortable sense of warmth when a systemic infection manifests itself with chills and cold hands and feet, as opposed to feverishness and elevated temperature.

Sages are very effective gargled for a sore throat and congested eustachian tubes and sinuses; an especially useful method is to gargle with a quarter cup of the tea, swallow it, and finish off with a menthol lozenge or some Eucalyptus or Desert Lavender tea.

All of them, especially the most aromatic species, are strong topical disinfectants, and, being both astringent and bacteriostatic, make good cleansing and analgesic washes for abrasions, contusions, and inflamed or chafed skin.

White Sage (*Salvia apiana*) was widely used as a soapless shampoo and conditioning rinse by Southern California Indians . . . it is indeed elegant.

OTHER USES: For cooking purposes, the best tasting of our native Sages are Black Sage (*S. mellifera*), Purple Sage (*S. leucophylla*), and Desert Sage (*S. dorrii*); the rest have their charm, to be sure, but not always in food. Bundles of white Sage branches are widely used for "smudging."

SAGEBRUSH

Artemisia tridentata   Compositae
OTHER NAMES: Chamiso, Chamiso
Hediondo, Big Sagebrush

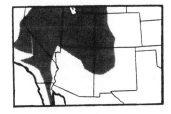

APPEARANCE: Yes, this is *the* Sagebrush, covering a large part of the high deserts of the western United States. Mature individuals are waist-high to chest-high, and colonies cover mile after mile with their grey-green drabness and pleasant scent. When they bloom in the fall and release pollen, many of us wheeze and sneeze and a few of us (mostly Native Americans) collect the pollen for ritual use. The bushes have grey-brown, flaky-shredded wood, the flowers crowd long branches in a slightly showy yellow display, forming the typical little-ball racemes of the Wormwoods. The silvery leaves are many, one-quarter to one inch long and three-toothed at the ends. The whole plant has that sagey, overpowering, and, if you are around it a lot, sickening scent typical of Sagebrush. Nearly all of the *Artemisias* (Wormwoods) are strong scented, but Sagebrush pulls out all the stops and overwhelms you with its no-nonsense proletariat stink.

HABITAT: All of the Great Basin, from southeastern British Columbia eastwards to the Dakotas, and south, down through the Mojave to nearly the sea in southern California.

COLLECTING AND PREPARATION: The leaves in their maximum spring and summer growth; best gathered from the heavy-foliaged bushes in favored areas along arroyos or roads. Dry, Method A, and remove the leaves afterwards. If the leaves are to be powdered for topical use, try to keep them in an airtight container in the freezer when not needed.

MEDICINAL USES: The powdered leaves are an old Native American remedy for diaper rash and moist-area chafing at any age; dust on the afflicted parts at regular intervals. If the irritation occurs in dry areas of the body instead, use a strong decoction made from the whole leaves and apply frequently also.

The whole plant is strongly antimicrobial and has a long history of usage as a first aid, disinfectant, and cleansing wash, from the prehistoric Anasazis to present-day Paiutes and Oregon farmers. In terms of frequency of use, Sagebrush is the high desert and Great Basin equivalent of Chaparral.

Sagebrush aromatics are also antimicrobial, and many people use or did use the smoke of burning leaves or the steam rising from moist Sagebrush on coals to clear the air (as it were) of pestilence and spirits of the dead. Even more than Juniper, Sagebrush is the old respected sauna and sweat lodge plant for ritual and literal purification.

OTHER USES: Fresh or dry, the leaves can be used as a cover and between for berries, seeds, or root foods, preserving them and protecting them from bugs and rodent attack.

I would be hard pressed to use them as the Utes used to, as an actual part of pemmican-like dried meat and berry food; the leaves taste quite dreadful.

I was once served turkey stuffed with "Sage" (actually Sagebrush) dressing, cooked up by some well-meaning folks from the city who were starting a farming commune in the most desolate land in northern New Mexico. The last time I saw them they were muttering about "navels of the Earth" and "negative Interfacing," eating Twinkies at a convenience store in Taos, on their way back to Berkeley. (Nobody was able to eat that turkey, by the way.)

SANGRE DE DRAGO

Jatropha cardiophylla, J. dioica, J. macrorhiza   Euphorbiaceae
OTHER NAMES: Sangre de Cristo, Dragon's Blood, Limber Bush, Tacote Prieto, Matácora, Leatherstem, *J. spathulata, J. cuneata*

APPEARANCE AND HABITAT: There are three distinctly different plants in appearance; so I'll discuss each one separately.

Jatropha cardiophylla is a medium-size shrub, usually found growing under small trees and taller bushes. It has flexible, rubbery branches and many bright-green, heart-shaped leaves. It is found in some abundance along dry slopes from Tucson to the west past Casa Grande, and south from Ajo through the Papago reservation and into Sonora and Baja California. It grows below 3,000 feet but above the driest valley floors.

Jatropha dioica is also called *J. spathulata* and *J. cuneata*; although some botanists feel they are separate species, this is, practically speaking, a single

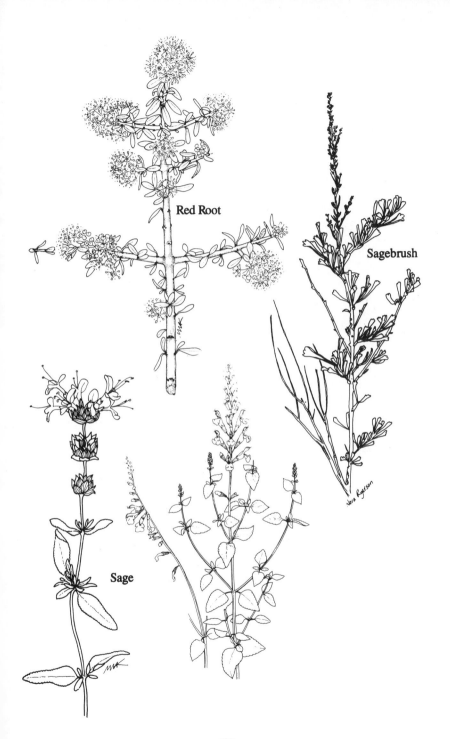

Red Root

Sagebrush

Sage

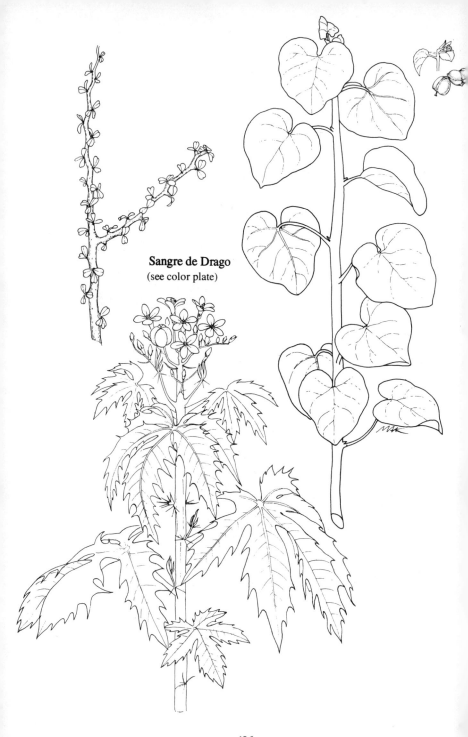

Sangre de Drago
(see color plate)

biotype. It is a tallish, many-branched shrub, usually five to eight feet tall, with thick and succulent stems and free-flowing, clear sap. The leaves are small and oblong, with two distinct growth patterns; on older, elephant-wrinkled branches they protrude from stubby little branchlets, but on younger branches they emerge directly from the bark. This Sangre de Drago grows in Arizona in the desolate low mountains of Yuma and Pima counties, often on hillsides below Elephant Trees; its habitat extends through Sonora, Chihuahua, and back into Texas in the Trans Pecos and Big Bend regions. It is found at 1,000 to 2,000 feet on dry hillsides and bajadas.

Jatropha macrorhiza, unlike the first two types, is a nonwoody herb. It puts up several large, coarsely toothed palmate leaves on long stalks, with rather showy white and yellow five-petaled flowers that form clusters at the ends of the stems. These mature into little, plump, green seedpods shaped like footballs. The root is enormous, an earth-tumor, warty and succulent-tough. The plant is found in abundance in the southeastern corner of Arizona and southwestern New Mexico, on mesas and rolling hills, from 3,500 to 7,000 feet.

CONSTITUENTS: Jatrophine, macranthine, jatrophatrione, jatropham (a lactam derivative), acetylaleuritolic acid, at least triterpene, oxalic acid, and phlebotannins.

COLLECTING AND PREPARATION: The hardened sap, branches, roots, and leaves of the first two species, for a decoction, chewing, or Method B tincture, 1:5, 50% alcohol. The root of the third, *Jatropha macrorhiza,* sliced while fresh, dried, Method B, powdered, tinctured, 1:5, 50% alcohol, in 20–40-drop doses.

MEDICINAL USES: The root of *J. macrorhiza* is a strong laxative, toxic in large doses, very similar in pharmacology and use to Cape Aloes and Jalap Root. In the small doses recommended, it stimulates colon peristalsis and decreases the time that intestinal contents can remain in the colon. This prevents, when appropriate, the contents from becoming over-dehydrated into hardened and constipative feces. The root is best used when fecal dryness is the result of body dehydration from sweating or lack of food volume, the result of extensive physical activity or fasting; it is not a particularly good laxative for the other, more common causes of constipation. Take the 20–40-drop dose with several glasses of water or electrolyte replacement in the early evening; the cathartic activity should be present by the following morning. Not for regular use, intestinal inflammations, or pregnancy.

The other Sangre de Dragos have nothing to do with this use; they are astringent, cooling, and topically anti-inflammatory. They may be taken as tincture or tea; useful in small frequent doses for diarrhea; applied topically or chewed for mild or severe gum or mouth sores, sore throat from external irritants or shouting, to soothe the stomach lining after nausea or vomiting; added to a sitz bath for inflamed hemorrhoids, vaginitis, or vulvitis, and applied to scrapes, cuts, and skin rashes to stop bleeding and reduce redness. This is as good a couple of plants for these purposes as any we have in our area.

Just remember, the *J. macrorhiza* is hot and irritating, active and expansive; *J. cardiophylla* and *J. dioica* are cold and soothing, shrinking and contracting.

OTHER USES: Widely used in Mexico to tan leather and dye it reddish-brown, *J. cardiophylla* and *J. dioica* also work on cloth, both as pigment (reddish pink) and a self-mordant.

SHEPHERD'S PURSE

Capsella bursa-pastoris   Cruciferae

OTHER NAMES: Thlaspi bursa-pastoris, Bursa, Bolsa de Pastor

APPEARANCE: A typical member of the Mustard family, with typical four-petaled flowers at the top of the stem, maturing into two-lobed, flat, heart-shaped seedpods filled with small, yellow-tan seeds. The basal rosette of leaves in the spring looks not unlike mustard greens but withers as the plant grows. Often only a foot tall, like any good annual weed it will sprout two leaves, a flower, go to seed, then grow its brains out the rest of the year. I have scythed my way through five-foot-tall stands of pure Shepherd's Purse (in front of the horse stables at Century Ranch in the Santa Monica Mountains) and pinched little six-inch bitties from creeks in March when I had run out of the tincture during the winter and had a friend with gout.

HABITAT: Widespread, common . . . and always hard to find when you want some. Starting in February and March (San Diego, Tucson, Austin) to late August (10,000 feet, below Mt. Graham, Arizona), this plant covers a wide range of growing conditions. Look for it in the spring around baseball diamonds in city parks and the vacant lots of Sunbelt subdivisions yet unsold. The remainder of the spring and summer it can be found around horses and cattle pastures, and around outhouses in the National Forests of our area.

CONSTITUENTS: Luteolin 7-rutinoside, quercetin 3-rutinoside, bursinic acid, fumaric acid and tyramine, choline.

COLLECTING: If it's green and alive, it's good, root and all.

PREPARATION: Not too durable after gathering (several months at the most), Method A or Method B, 1:5, 50% alcohol. If you prefer the tea, dry the herb carefully and freeze it in airtight bags until needed.

MEDICINAL USES: Shepherd's Purse has four main uses: 1) to stop bleeding under certain conditions, 2) to relieve inflammation in acute urinary infection, 3) to stimulate kidney excretion of uric acid, and 4) to strengthen and synergize the effects of native oxytocin in homebirths. So, if you have gout or are a midwife, you need this stuff.

Now as to the hemostatic effects. The rubinosides are capillary strengthening, and the bursinic acid (and perhaps the tyramine) is vasoconstricting; so use Shepherd's Purse when you have heavy menses that are bright red after several days and not turning brown, or whenever there is midcycle bleeding, either from stress, drugs, or fibroids (fibromyomas and the like). Some docs will say that all of those symptoms are only for medical treatment; actually, many are sub-pathologies and, if push comes to shove, the doc will say they are untreatable, or "If they don't get better, come back for another pelvic in three months and we'll do a D&C on/for you." In the meantime, take some Shepherd's Purse, okay? . . . once you know nothing is seriously wrong, that is. Bleeding can be nothing much or very serious, whatever the tissue involved; so always have a medical review before self-treatment. Blood in the urine after an infection, rectal bleeding, coughing of a little blood after a week of bronchitis; they all may be short-term, mild problems or long-term, serious ones. If they are medically mild, use Shepherd's Purse. If the bleeding doesn't stop in a couple of days, go back to the doc again.

For the basic attack of urethritis or cystitis, from food binges, too much booze or coffee, that strange white powder you snuffed a lot of at a party two nights ago, rabid and enjoyable sessions of sexual excesses that leave you walking bowlegged, or driving nonstop from Santa Barbara to Bangor, Maine, with only a bag of dried banana chips, two pounds of Nacho Flavored Fritos, and homemade salsa for food, in a rented subcompact with twelve traveling bags, a parakeet, and a four-year-old, cranky child . . . your basic garden variety of bladder infection. Drink ½ teaspoon of the tincture (fresh or dry plant) in a cup of warm water every three hours, lay off the carbos or burgers, and get over it. The more acute the onset, the better it works. Sluggish, week-long infections usually need stronger medicine. Recurring short-term infections point to diet excesses or bad stress habits and this little herb will have only a palliative effect.

As for the uric acid effects, Shepherd's Purse helps to facilitate more efficient excretion by the kidneys of this nucleoprotein waste-product. Decreasing blood levels when they are elevated (hyperuricemia) decrease the frequency of gout and pseudogout attacks as well as the severity and frequency of arthritis episodes. This is even more pronounced when there are episodes of phosphaturia not directly the result of kidney damage or parathyroid imbalances, as Shepherd's Purse helps stimulate better phosphate recycling by the kidneys. Although the mechanism is not understood, the effects of oxytocin on the uterine lining are increased when the herb, alone or with our other oxytocin synergist, Cotton Root, is given during birth. Try 1 teaspoon of the tincture in warm water, a strong, warm (not hot) cup of the tea, or ½ teaspoon Shepherd's Purse tincture with 1 teaspoon Cotton Root Bark tincture in warm water, sipped slowly, *after* most of the cervical dilation has occurred. At least twice that I know of, midwives or physicians using Shepherd's Purse before substantial dilation have seen hourglass contractions occur in the mother's uterus.

CONTRAINDICATIONS: "Staghorn" uric-acid stones in the pelvis of the kidneys may be partially dislodged and irritate the ureters; the effects of the vasoconstriction may aggravate labile hypertension in some individuals, particularly the aged; not for use in overt kidney disease; may stimulate uterine contractions if used during pregnancy.

SILK TASSEL

*Garrya flavescens, G. wrightii, G.
goldmanii*   Garryaceae (Cornaceae)
OTHER NAMES: Quinine Bush, Guachichi,
Cuauchichic, Bear Brush

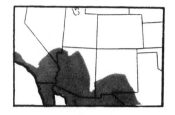

APPEARANCE: These are evergreen shrubs, resembling Oaks or Manzanitas a bit, with stout, uncomplicated oval leaves, opposite and short stemmed, usually one and one-half inches long, and pointed. The shrubs of all of these species are grey-green, with *G. flavescens* (Yellow Silk-Tassel) having distinctly yellow-faded older leaves. The flowers are yellow to light-cream-colored, separate-sexed, and droop down in catkins, sometimes spectacularly. The seeds are little, blackish, dry berries, a quarter inch around, borne at the ends of the branches. Every part of the plant tastes terrible. The bushes range up to ten feet in height and are substantial in demeanor.

HABITAT: From 3,000 to 7,000 feet in our drier mountains. You usually find Silk Tassel growing just below or in the lower sections of the Juniper/Piñon belt, whatever that altitude is in your area. The major coastal California species, *Garrya fremontii*, makes it south into some of the drier areas of the Santa Ana Mountains and interbreeds freely with Yellow Silk Tassel in the southern Sierra Nevada and the higher mountains of the Mojave. *All* of the Garryas interbreed; so distinct special differences may be somewhat suspect.

CONSTITUENTS: Garryine, garryfoline, veatchine, delphinine, cuauchichicine, and two unnamed diterpene alkaloids.

COLLECTING: Since Silk Tassel can be toxic, the ratio of strength is approximately thus: leaves 1, bark 4, root and bark 10. You can get the same results from any part of the plant, but adjust dosage accordingly. The leaves are best in summer, the bark and root in fall; dry however is appropriate.

PREPARATION: Tincture, Method A, 1:5, 50% alcohol. The dosage ranges from ½ teaspoon of the leaf tincture five times a day to 10–15 drops of the root tincture five times a day.

MEDICINAL USES: Silk Tassel is a strong and reliable smooth-muscle relaxant, one of those generally classed as parasympathetic inhibitors or anticholinergics. In proper doses it has little effect on the central nervous system but slows down the impulses of the vagus nerve, myenteric plexus nerves, and the sacral ganglia of the parasympathetics. This makes it a useful pain reliever and antispasmodic for diarrhea, dysentery, gallbladder attacks, and menstrual cramps. Mojave and Kawaiisu Indians used it for stomach cramps and diarrhea, and it works very nicely for the cramps of gas and flatus, both upwards- and downwards-striving. I have seen it very helpful in small doses for the distress of hiatus hernia gas pain. It has certain advantages over classic anticholinergic medicines, such as the atropine type, because it has little secretory suppression and won't cause the dry-mouth and constipation syndrome.

CONTRAINDICATIONS: Prescription medication and over-the-counter drugs should be avoided when using Silk Tassel, and it is not appropriate during pregnancy or for children. If you start to get short of breath and cold-clammy, you have taken too much (hard to do, fortunately).

SILVER SAGE

Artemisia frigida   Compositae
OTHER NAMES: Estafiate, Fringed Sagebrush, Romerillo, Istafiate

APPEARANCE: Not a true Sage (*Salvia* spp.), this is a little Wormwood, up to a foot high, with a brittle, raggedy, and somewhat woody base and stems. It is covered in many finely divided aromatic leaves, altogether forming a feathery, silver-grey mass. In the summer it sends up many small, slender, typical Wormwood stalks. In our overgrazed high and cold deserts it is often hugely abundant, uneaten by cattle and with little surviving plant competition. In the very high mountains False Hellebore can dominate the overgrazed meadows, below that Yarrow, lower still, Sagebrush, then Silver Sage, and finally Escoba de la Vibora, all attesting to our national craving for cheeseburgers.

HABITAT: 4,000 to 8,500 feet in the Great Basin, east to Colorado, south to Arizona, New Mexico, and the Panhandle. Look for Juniper, Piñon, lots of cattle, and little else.

CONSTITUENTS: Various lactones, guaianolides, 8-deoxycumambrin.

COLLECTING: The leaves and flower stalks, Method A or B to dry, depending on the size of the plant. Gather from July to September for the best value.

PREPARATION: Tincture Method B, 1:5, 50% alcohol, hot infusion, cold infusion.

MEDICINAL USES: The tincture, from 20–30 drops in a little cold water, effectively retards stomach hypersecretions, for those who oversecrete between meals or wake up at night with acid indigestion. If this is a major, long-term problem, try the tincture followed by some Alfalfa or Red Clover tea in midafternoon, before retiring, and in place of coffee or fruit juice first thing in the morning.

The simple tea, brewed hot from a generous pinch of the fluffy herb, is a strong diuretic and moderate laxative for "dry" individuals and has been a major "hot" herb in Tewa folk medicine for centuries. Like California Mugwort, the tea makes a useful forehead and scalp wash for frontal headaches; even better is a vinegar tincture, Method A, 1:5, applied on the sides of the face and temples when there is the headache accompanied by bloodshot or red eyes; it smells weird, but it works.

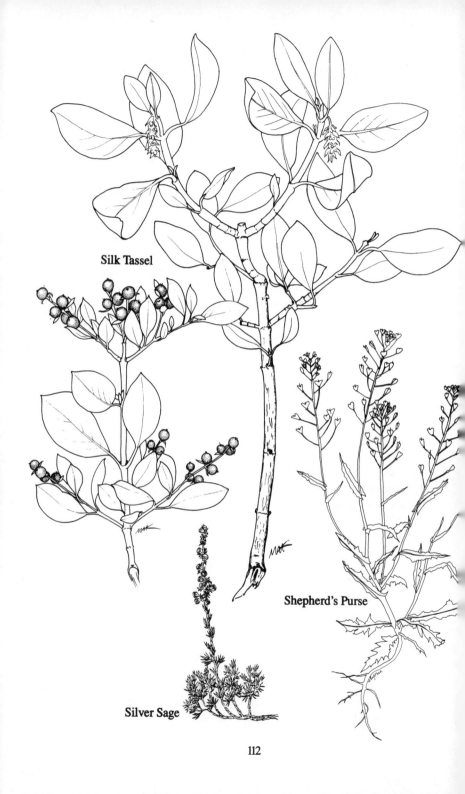

Silk Tassel

Shepherd's Purse

Silver Sage

112

SOAPBERRY

Sapindus saponaria, var. *drummondii*
Sapindaceae

OTHER NAMES: *S. drummondii,* Cherioni,
Jaboncillo, Palo Blanco, Wild Chinaberry
(usually applied to *Melia azaderach,* or
Pride-of-India)

APPEARANCE: This is a small tree or large bush, usually six to fifteen feet tall in our area, higher farther east and in central Mexico. The leaves are large, languid, and pinnate; each of the nine to nineteen leaflets is two or three inches long, lanceolate, and smooth edged. The twigs and small branches are yellow-green and downy, the bark of the one or several trunks is a scaly-smooth, yellow-grey or tan color. The many tiny flowers bloom in late spring and form large, dense, intensely fragrant terminal clusters. These mature into strange, amber-bead, translucent yellow fruit, about one-half inch in diameter, with a single seed inside. The sticky fruits persist well into the following spring, gradually shrinking. The whole tree has an elegant and somewhat decorative appearance.

HABITAT: Arizona, from Cottonwood, along the Verde, Salt River, and Gila drainages, south into Mexico, through all of southern New Mexico, and eastwards to southern Missouri and western Arkansas, from 2,500 to 6,000 feet. Its preferred growing conditions are very unpredictable; it likes hills, valleys, grasslands, oak woodlands, even mountainsides. Fortunately, those sticky-soapy, amber-colored bead-berries give it away.

CONSTITUENTS: Fruit: 20% fixed oil, a-b-amyrins, B-sitosterol, sanguinarine; leaves and stems: luteolin, rutin, 4'-methoxy flavone, sanguinarine.

COLLECTING AND PREPARATION: The leaves and stems, dried, Method A, gathered in summer or early fall. The fruit should be gathered when recently ripe, usually August or September, sometimes later. The leaves and stems may be chopped when dry and used as a cold Standard Infusion, 2–4 ounces up to three times a day. The berries, for soap: simply crush the fresh fruit and lather away, or juice or puree them, strain out the particles, and preserve them with alcohol: 2 parts juice, 1 part 95% grain alcohol.

STABILITY: Leaves and twigs, indefinitely; fresh berries may be refrigerated for a month or more.

MEDICINAL USES: The herb tea is a distinct anti-inflammatory and analgesic. It is in fact one of our better arthritis remedies for internal use. Use it for acute episodes, not for chronic arthritis or during remissions. It is not well suited for constant use, like Cottonwood, but may be rotated with Yucca or Agave.

On those occasions when, for whatever reasons, you feel sick and sluggish and you have a moderate fever that persists for several evenings in a row, with sweating and physical agitation, then wake up in the morning, feel fine, get hot again that evening (what the old-time docs called periodic fever), try the Cold

Infusion, warmed up to hot and drunk in the late evening. Take it alone or in combinations with Indian Root tincture, Silk Tassel tincture, and Echinacea. See Formula #14 (page 139).

The berries, freshly crushed, make an excellent if somewhat messy laundry soap. The preserved juice is a good hand and hair soap. It has medicinal effects as a soap also, helping to relieve the itching and prevent the spread of various tineas and scalp seborrhea. As it may cause a mild form of dermatitis in a few individuals, it is best to test a little first. Place a few drops on the inside of the wrist and wait a few hours. If no reaction occurs, it is safe to use.

OTHER USES: It will poison fish without altering their taste, but, as with Turkey Mullein, how do you get them to stay in one place long enough?

SPIKENARD

Aralia racemosa, A. humilis   Araliaceae
OTHER NAMES: *A. bicrenata, A. arizonica,*
Elk Clover, American Spikenard, Spignet

APPEARANCE: These are large, robust plants, often six to eight feet tall, with large, graceful compound, usually pinnate leaves, with serrated little teeth all along the leaflets . . . in our range at least. The greenish white flowers form umbels like plants in the Parsley (Umbelliferae) family; the little berries are purple and sweet-spicey like those of its close relative, Ginseng. Also like Ginseng, the older plants show distinct, step-like leaf scars at the top of the root, each scar showing a previous year's growth. The root is cream colored, brown skinned, and fleshy, with oil oozing from cuts, especially around the cortex. The roots may be six feet long and form massive tangles of arms, legs, and other anthropomorphic detritus. Unlike botanically close relatives in the Umbelliferae such as Angelica, Osha, and Cow Parsnip, the stems of Spikenard are solid, not hollow. Botanists cannot decide whether the Umbelliferae evolved from the Araliaceae or vice versa, but the large varieties of both families are obviously related.

HABITAT: In Arizona and New Mexico, Spikenard grows up in the wettest, coolest, and shadiest canyons of the desert mountains, usually from 4,500 to 7,500 feet. If you are hiking about in the low, hot foothills, saving your break for that year-round stream up the canyon over there (it looks fine on your topo map) . . . going to take a lunch break, maybe swim in a pool some friend told you about (after all, it drains that wilderness area up on top), climb a little farther up that shady canyon into the boulder rubble, past moss and monkeyflowers (ignore for the moment those tempting pools), and you will probably come across a stand of Spikenard glowering hugely in the darkest crevices. Dig a length or two of the root, maybe a whole smaller plant downstream from the big ones, shade yourself in the great green umbrella of the plants, chew some berries if there are still any on them, marvel at your/our cleverness . . . *then* go for that swim lower down in the canyon pool. Now do you understand why some of us are herbalists?

CONSTITUENTS: Choline, chlorogenic acid, ursolic acids, b-sitosterol, araloside, oleanic acid glycosides, and several panaxosides.

COLLECTING: The root, from midsummer to winter. Large root masses need to be split and separated from rotten or dead sections. Dry the roots in reasonably intact pieces; break down for processing as needed.

STABILITY: Dry roots are good for about a year.

PREPARATION: Tincture dry, Method B, 1:5, 60% alcohol; for a honey extract, use 1 part fresh root in 5 parts honey, finely chop the root and simmer slowly in the honey for about an hour, strain, and bottle. The tea is fine, either Standard Infusion or a Weak Decoction, but always chop the larger, dried pieces down to tea size when needed, to keep the rather delicate aromatics as much intact as possible.

MEDICINAL USES: Like other close relatives of Ginseng, Spikenard has shown an ability to stimulate phagocytosis in white blood cells, increase interferon synthesis in infected cells, and increase the capacity for metabolic stress in rats. (I haven't done too much counselling with rats, but I can vouch for its helping human beings.) This function of Spikenard is sometimes adaptogenic, increasing mobilization but decreasing the metabolic costs of stress responses. This *may* mean (the jury is still out) that moderate amounts of the tincture or tea on a regular basis can strengthen someone with metabolic or chronic disease, whatever the type.

More prosaic but more predictable, Spikenard is a first-class medicine for the initial stages of bronchitis, pneumonia, bronchorrhea . . . all that stuff we usually call a "chest cold." The tincture (¼ – ½ teaspoon in hot water), the tea (2–4 ounces hot), or the honey cough syrup described above (1–2 teaspoons) works well for adult or child. Conversely, the same amounts will help the individual with moist, tired, chronic coughing; the aged person with impaired pulmonary function; or the heavy smoker or former smoker with a moist, phlegmy cough in the mornings and evenings. For this latter group, the more the sense of chest and lung tiredness, the better Spikenard works.

The Pomo Indians of California and Mescalero Apaches both used it as a strong boiled tea for "Winter Chest," as well as for frequent stomachache and morning nausea. Other traditional uses include making a strong, dark tea or using the powdered root and applying either to open, itching sores and eruptions.

A hot tea of the root will usually help start menstruation when the month has been a hard one, with a head cold or sudden change of weather possibly delaying the onset. Although not as pronounced in effect as Inmortal, Spikenard (several cups a day) helps maintain coherent lochia flow after birthing. It has a lot of supporters amongst midwives in this country and Canada, being used along with Squaw Vine (*Mitchella repens*) and Raspberry leaf tea as a prepartum uterine tonic for the last few weeks of pregnancy. As it is impossible to do a control-group study (let alone a double-blind one) on a single pregnancy, all such information must be deemed as "anecdotal." Nonetheless, for two hundred years of white and black midwifery, and who knows how many years of Native American usage, it has been "anecdoted" to help for an easier birth.

OTHER USES: It is a standard ingredient in traditional root beer recipes. Although Spikenard has nothing to do with the biblical Spikenard, the outer bark of especially aromatic roots from the plant does make a pleasant incense.

STILLINGIA

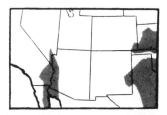

Stillingia sylvatica and others   Euphorbaceae

OTHER NAMES: Queens Root, Yaw Root, Queen's Delight, Pavil

APPEARANCE AND HABITAT: We have two major groups of Stillingias in our deserts; one is a group of closely related species found in Texas and eastern New Mexico, the other is found in the Colorado Desert and the deserts of Arizona. The eastern group includes *Stillingia sylvatica* and *S. texana*. These two are plants of sandy, high flood plains and open limestone gravel, from Arkansas west to Roswell, New Mexico (especially in the Pecos and Canadian River basins). They are two to three feet tall, many-branched, and emit caustic, milky sap when broken. The leaves are thick, stem-clasping, and elongate; the flowers are yellow, located at the tips of the flowering branches; the round, Poinsettia-like, three-lobed capsules grow right below the flowers, until all are capsules. The roots are long, fleshy, and tapered, reddish brown and pink-brown inside, often forming dozens of snakelike, seldom-branching masses, especially in deep, sandy soil. These two are the preferred species.

In the Colorado Desert and the deserts of Arizona live *S. paucidentata* and *S. linearifolia,* the first with growth similar to that of *S. sylvatica* and *S. texana*, but they are usually under a foot high, the flowers are mauve-purple to dark green, and the root is usually single and deep; the whole plant smells rather bad. It is found in the lowest areas of the western Mojave Desert to below sea level near Mecca, California. *S. linearifolia* grows in washes and on the edges of valleys in the southern Mojave, most of the Colorado Desert, a little around Yuma, Arizona; it has many slender stems, is one to one and one-half feet tall, thinly leafed by slender willowy leaves and yellow-green flowers, small, sparsely placed on the stems, the flowering-fruiting stems usually extending above the leafed stems. The root is also usually single, thicker, and woody, and not as succulent or as deep.

CONSTITUENTS: Stillingine, echiine, apocinide, sylvacrol, volatile oils.

COLLECTING: The roots, sliced lengthwise into quarters while fresh, the strips cut into small sections after dried and before tincturing. However the roots are processed they should be extracted before they are more than six months old, as

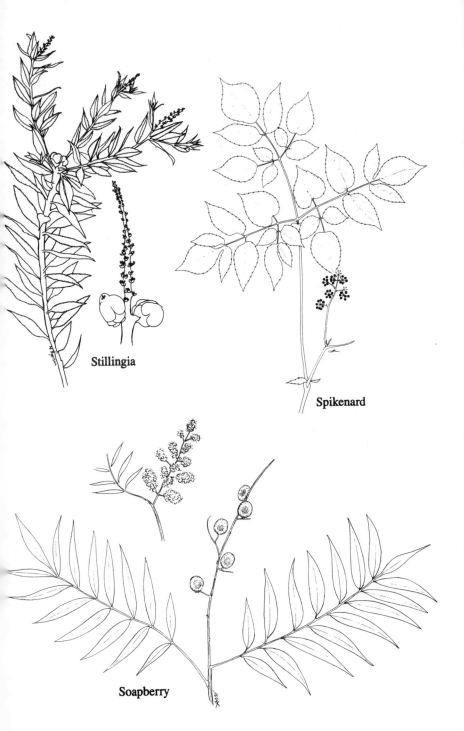

Stillingia

Spikenard

Soapberry

117

they deteriorate rapidly.

PREPARATION: Tincture, Method A, of dried root, 1:5, 50% alcohol. A lovely alterative syrup may be made from Stillingia and some of the other plants in this book; see Formula #15 (page tk) for the recipe. This may be used, 1–2 teaspoons a day for arthritis, eczema, psoriasis, rheumatoid arthritis, and other symptoms of chronic allergies and mild autoimmune conditions.

MEDICINAL USES: Besides the above formula and its use, Stillingia tincture has a strong effect on subcutaneous and submucosa fluid retention and impaired lymph drainage from those areas. It and the syrup work well for chronic bronchitis, smokers' cough, and protracted recuperation from lung and bronchial conditions. Moist skin eruptions such as eczema and dermatitis respond well to Stillingia, especially when the lymph nodes that drain the irritated skin are slightly enlarged. Long-standing tineas and fungal infections may sometimes be helped, especially when topical treatment with such herbs as Tronadora, Desert Willow, or Cypress has not helped or if pharmaceutical ointments have had little effect. The morning cough and dry, raw throat that some of us get in the winter, especially when the air is cold and dry, usually goes away when a little tincture or syrup is taken. The tincture should be used in 20–40-drop doses.

SIDE EFFECTS: Too much Stillingia can cause nausea and loose stools; so use it conservatively.

SUMACH

Rhus glabra, R. trilobata, R. microphylla, R. ovata, etc.　　　　Anarcadaceae

OTHER NAMES: Sumac, Smooth Sumach, Sugar Sumach, Lemonadeberry, Lemita, Squaw Bush, Pajul del Norte

APPEARANCE: We have a lot of Sumachs in the West; so let's start with the number of leaves. *Rhus ovata* and its close relative *R. kearneyi,* called Sugar Sumach and Kearneys Sumach, respectively, form large handsome bushes, usually from four to ten feet tall, with large, oval, single leaves, bluish-green, leathery, and with a deep central vein. Both of these Sumachs form terminal cream-white puffs of flowers, maturing into the sticky-hairy flat red berries typical of the whole group. The one Sumach with three leaves is the Squawbush (or Skunkbush), *R. trilobata,* and looks suspiciously like a miniature-leaved Poison Oak bush. Poison Oak/Ivy is a relative of all the Sumachs and has the botanical names of *R. toxicodendron, R. radicans, R. diversiloba, Toxicodendron radicans, T. diversiloba, T. rydbergia,* etc. Our Poison Ivy/Oak is a large, three-leaved, low-growing semivine that forms *white* berries, never sticky-hairy red, so don't worry . . . too much. The Squawbush (Skunkbush) is a tidy, much varied bush/shrub with a slightly unpleasant smell and, in the summer and fall, many little tight clusters of five or ten berries.

The pinnate-leaved Sumachs have milky sap (usually), and, starting with those with five leaves, we have Mearns Sumach and Evergreen Sumach (*R. choriophylla* and *R. virens*), medium-large shrubs, symmetrical and attractive, with the usual cream-white flowers forming sticky-hairy red berry clusters; we have Little-leaf Sumach (*R. microphylla*), a dark and snaggly Mesquite-like bush that greens out in summer to form first the flowers, then the seven to eleven little pinnate leaflets, then the berries; finally, we have the classic mainstream Sumach, with many pinnate leaves, often forming the thickets people from the eastern states think of . . . Smooth Sumach and Prairie Sumach (*R. glabra* and *R. copallina*), forming stands of waist- to head-high plants with walnut-like compound leaves of up to thirty-one to thirty-three leaves, forming dense clusters of white flowers, and then dense, even top-heavy clusters of brick-red berries. The leaves turn from a shocking blood-red to purple-red color in the fall.

HABITAT: Sugar Sumach is found in central Arizona, all along the Mogollon Rim, and in California in much of the desert mountains of the southern counties on to the coast, as far north as Santa Barbara, usually from 2,000 to 4,000 feet. Kearneys Sumach is found in the same area of Yuma County and Sonora as Chaparro Amargosa. Squawbush is everywhere, interspersed in the Great Basin of Utah, Idaho, and Northern Nevada with Utah Squawbush, a similar plant with single, odd-shaped leaves. Mearns Sumach is a middle and lower mountain plant that grows from the foothills of the Catalinas near Tucson to the foothills of the Sandias south of Albuquerque, and Evergreen Sumach is a similar foothill bush of the Guadalupe and Davis mountains of New Mexico and Texas. Littleleaf Sumach grows from the Huachucas to Guadalupe Canyon in southeastern Arizona and the Guadalupes in southern New Mexico, West Texas, Sonora and Chihuahua states, from 3,500 to 6,000 feet in gravelly soil and open, overgrazed (former) grasslands. Smooth Sumach grows in most of the moist canyonlands of Arizona and New Mexico, from 4,500 to 7,000 feet, north in warm canyons of Colorado, Utah to Canada, east to the Atlantic Ocean. Prairie Sumach is found in the Guadalupe and Davis mountains of New Mexico and Texas and sporadically eastwards. A Sumach for every clime.

CONSTITUENTS: The fruits contain calcium and potassium malates, tannins, fixed and volatile oils; the dried leaves and bark contain gallic and tannic acids and various polyphenols.

COLLECTING: Gather the leaving stems when green, and dry (depending on size of the plants), Method A or B. Gather the berries when they are fully red but while the foliage is green or moderately red; dry the berries in intact bunches, Method B. For making "lemonade" to drink (properly, Rhusade), steep a rounded tablespoon of the fresh or dried berries in a cup of water until tasty.

STABILITY: The leaves a year, the berries two or three years.

PREPARATION: The leaves can be powdered, sifted, and stored for topical use. For a quick salve, thoroughly stir the powdered leaves (part by volume) into 2 parts (by volume) of petroleum jelly, such as Vaseline. For a glycerine tincture, using Method B, macerate 1 part (by weight) of the powdered leaves in a mixture of 5 parts of a menstruum that is half glycerine and half water, leaving it to macerate for longer than usual, say four weeks. Agitate and strain as usual.

MEDICINAL USES: The powdered leaves, quick salve, or the glycerine tincture are excellent treatments for all types of muco-epithelial sores, from fissures to ulcers to mechanical injuries, on the lips, mouth membranes, genitals, or nostril membranes, acting to soothe, shrink, and mildly disinfect. The powdered leaves are very soothing to mouth sores in nursing infants, and the tincture can be diluted with twice as much warm water as a nasal spray for nostril soreness from smoke, dust, or snuff or cocaine abuse.

OTHER USES: As mentioned under Preparation, the steeped berries, sweetened or unsweetened, make a very cooling and refrigerant drink in hot weather. In parts of California and Texas I have heard the name lemonadeberry applied to several Sumachs, and academic types sometimes prefer the more technical name, Rhusade, for the nifty little iced drink made thereof.

SYRIAN RUE

Peganum harmala   Zygophylaceae
OTHER NAMES: African Rue, "Soma"

APPEARANCE: Syrian Rue is a bright green, smooth-surfaced roadside weed, composed of many one-and-one-half- to two-and-one-half-foot-tall basal branches, divided further into smaller ones and finally dividing into shiny, somewhat succulent, thready leaves. The foliage has a strong, subrank scent, and through most of the spring, summer, and fall it maintains pretty, little five-petaled flowers growing in the leaf axils. These mature into round hollow capsules, which often survive into the following year on the underside of dead branches fringed around the base under the new growth. They contain many small, angular seeds; brown capsules are the recent products, grey capsules are from previous years. The root is grey-brown, pithy, with a yellow heartwood . . . and hard to dig up.

HABITAT: A native, like Tumbleweed, of northern India, Afghanistan, and southern Russia, it was first noticed growing in the United States in the 1930s near Fallon, Nevada and Deming, New Mexico. I have seen it increasing its presence here, now growing from El Paso north to the Guadalupe Mountains, westward through the southern counties of New Mexico into Arizona. It is most frequently seen along secondary paved and dirt roads in our lower canyons, alluvial flats, and grazing lands. Because the plant is especially poisonous to sheep, eradication projects have continued over the years, with some success. The only problem is that the rural highways get regular maintenance and mowing; the capsules are carried for miles, caught between the blades of machinery, and sown gleefully (for the plant) and tragically (for the sheep) farther and farther away from the original plants.

CONSTITUENTS: Seeds: harmaline, peganine, harmine, vasicinone; root: same, plus harmol; herb: peganine, vasicinone, and 5,6-hydroxytryptamine, peganidine, and deoxypeganine. All or parts of these constituents make the plant taste truly gross.

COLLECTING: The seeds as whole capsules, gathered in the fall, winter, or early spring; the most recent ones are the best. The root is strongest in the late fall, when most of the leaves are dead. Dry the herb, Method A, picked in mid-summer.

STABILITY: Root and seeds last for years; foliage, no longer than a year.

PREPARATIONS: The seeds, tinctured, Method B, 1:5, 50% alcohol, or ground and encapsulated. The root, dry, tincture, Method B, 1:5, 60% alcohol. The dried herb as a tea, Standard Infusion, or as a tincture, Method B, 1:5, 50% alcohol.

MEDICINAL USES: Recent Russian studies have verified the effectiveness of many of the folk usages in Russia and India, particularly its treatment for skin conditions. Most types of contact, eczematous, and exfoliative dermatitis and psoriasis respond well to an external wash of the herb or root tea and internal use of the tincture or tea in small frequent doses . . . 10–20 drops four or five times a day during acute episodes. The herb tea is also an excellent hair and scalp rinse for chronic seborrheic dermatitis (dandruff); rinse the scalp with the tea after shampooing and rub it in vigorously. A little bit brushed into the scalp daily (the tea or tincture, not the dried plant) is still a common practice for treating dandruff in India. The hair can get a bit stiff from this, at least for some people, so you may have to work with a mixed bag of cosmetic and herbal treatment to juggle both the dandruff and the visual aesthetics.

The seeds (tincture, 40 drops, or capsule, 1 #00) are a useful antidepressant and mood elevator for folks with the mopey dragass depressions, not the nervous, peripatetic, manic depressions. People who sit in front of the television all day long (whether or not they turn it on) and don't want to go out or be visited usually find that Syrian Rue and a noisy friend can shake them out of their malaise.

There are clearly defined cardiovascular effects from the plant, especially from the seeds. Moderate doses (40–50 drops) increase the force of the pulse, peak aortal flow, and myocardial contractile force but decrease the pulse rate and total blood pressure; this may be part of its mood-elevating effect and also part of the reason many old people in the eastern states of the U.S.S.R. take the tea regularly. Besides, it probably grows outside their homes.

I once was approached by a professor from California looking for plant specimens and gathering localities; he was sure that Soma, the near-mythic plant drug mentioned in the early Sanskrit holy works, was actually Syrian Rue and not, as generally accepted, the rather poisonous mushroom *Amanita muscaria*. True or not, for all you fellow aging hippies out there, ingesting a dozen capsules filled with the seeds or taking some of the root as a (gag) tea certainly alters your consciousness. Of course, so will lying sick in bed with a 103° fever or chugalugging ouzo.

OTHER USES: (I personally never liked sheep) . . .

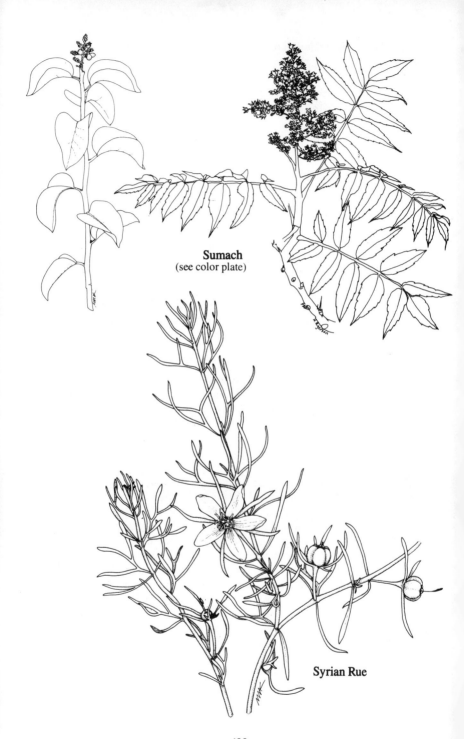

Sumach
(see color plate)

Syrian Rue

122

TRONADORA

Tecoma stans   Bigniniaceae
OTHER NAMES: *Stenolobium stans, S. incisum*, Trumpetflower, Esperanza, Palo de Arco, Yellow Elder, Yellow Trumpet

APPEARANCE: This is an attractive shrub, from three to six feet high, broad, a little unkempt and straggly; some branches extend up, some sideways, some droop downward. The opposite leaves are pinnate, with seven to thirteen leaflets, each one to two inches long, lanceolate, and regularly, neatly toothed. The foliage is a striking, bright, shiny green. The flowers are even more interesting, trumpet-shaped and up to two inches long, blossoming in terminal yellow masses from the branches.

HABITAT: Southeastern Arizona, southwestern New Mexico, western and southwestern Texas, deep into Mexico. It is fond of slopes in the more substantial desert mountains, especially the Oak woodlands with grasses and loose gravelly soil, from 3,000 to 5,500 feet.

CONSTITUENTS: Various piperidine-type monoterpinoid alkaloids; tecostanine, tecomanine, boschniakine, actinidine, a methyl-ester of 4-methoxy-transcinnamic acid, a type of polyphenol oxidase, and an enzyme that reduces catechol to 3,4,3',4'-tetrahydroxydiphenyl.

COLLECTING AND PREPARATION: The leaves and flowers, dried, Method A. As they are rather stable, they can be tinctured, Method B, 1:5, 50% alcohol, powdered for encapsulating; or as a Standard Infusion, 3–4 ounces at a time. The wild plant is distinctly more potent than cultivated ones.

MEDICINAL USES: Tronadora is widely used in Mexico and Central America as a treatment for adult-onset insulin-resistant diabetes. Unlike Matarique or Prodigiosa, which decrease gluconeogenesis, the formation and release of glucose from stored nutrients by the liver, Tronadora apparently stimulates liver glycogenesis, the formation of complex, stored nutrients from blood glucose. In other words, since the constant elevation of blood sugar is the problem, and the liver is the source of that blood sugar, Tronadora stimulates the return by the liver of the glucose into storage glycogen; the other two herbs block the release by the liver of the excessive glucose. It has been used for many years south of the border, both by professionals and by regular folks. One thing is widely accepted when using Tronadora; the person taking it must decrease his or her caloric intake 5 to 10 percent for the herb to work. This is almost always necessary for such diabetics, anyway, but the pharmacology of the plant seems to be such that the decreased calories cause the herb to stimulate the liver.

Most antidiabetic medicines have substantial toxic potential; if Tronadora helps, then it has distinct advantages over the prescription approaches because it has virtually no toxicity. In Latin American medicine, Tronadora goes in and out of fashion, like gold preparations for arthritis, which go in and out of use about every ten years. Tronadora, in various pharmaceutical forms, is part of standard practice; then it gets declared useless; then comes back again. The poor folks keep on using it, in or out of favor, with or without the approval of the doctors. The Mexican pharmacopoeia recommends a dose of 2-10 grams a day, with 5-6 grams being the average. Ten to twelve capsules a day of the powdered herb or two 4-ounce doses of the standard infusion seem like a reasonable regimen.

The tea is also very useful for the gastritis suffered by heavy drinkers or those of us who do binge drinking; half a cup of the tea in the morning or 1 teaspoon of the tincture in some warm water works well. The herb is antiviral, particularly against adenoviruses and herpes simplex virus, so take it as a tincture or tea when you have a head cold coming on, alone or with Echinacea.

TRUMPET CREEPER

Campsis radicans   Bignoniaceae

OTHER NAMES: *Tecoma radicans, Bignonia radicans*, Cowitch

APPEARANCE: Trumpet Creeper presents you with a barrage of dark green leaves and impressive reddish orange or scarlet flowers crawling all over the place, up old mine shafts, down telephone poles. The leaves are shiny and ornately compounded, branching from opposite sides of the vine stem; the flowers, improbably virulent, erupt in bunches of three-inch-long trumpety blossoms.

HABITAT: A native plant of the Southeast, it grows naturally as far west as central Texas. Although listed in few references regarding the flora of our western states, it is a common urban adventurer in some of the older mining towns and semighost towns of the West. Brought in as a cultivator by early settlers, I have seen it scrambling over old shamble-shacks in such off-the-wall places as Searchlight, Nevada; Bisbee, Arizona; and ruins of old mines in the Providence Mountains of California and around crumbling adobes in central New Mexico. It is definitely well established and naturalized in many localities, even if, like wild Eucalyptus trees, botanists haven't noticed it yet.

CONSTITUENTS: Alkanes, squalene, salicylic acid, 3,4,5-trimethoxycinnamic acid, and boschniakine, a monoterpene pyridine alkaloid; the flowers contain cyanidin-3-rutinoside.

COLLECTING: The whole vine in flower, dried, Method A.

PREPARATION: The whole plant, powdered for tea or a vinegar tincture, 1:5.

MEDICINAL USES: The herb is used as a douche for Candida infections and, to a lesser degree, concurrent hemophilus infection. Use 4 tablespoons of the vinegar tincture in a pint of warm, isotonic water (½ teaspoon of salt to a pint of water). You might add a teaspoon of Yerba Mansa or Echinacea tincture as well, particularly if it has lingered for awhile. Use the douche every other day for a week or two.

The tea has a number of traditional uses for such things as barber's itch, farmer's nails, and athlete's foot . . . or is it barber's nails and farmer's foot . . . anyway, for various skin funguses and tineas you might try the tea or use the powdered herb as a simple dust, alone or combined with Desert Willow, Chaparral, Arizona Walnut, or Cypress.

SIDE EFFECTS: The fresh plant may cause a contact dermatitis for some; so if your skin is sensitive, use gloves when gathering. It's fine after it is dried.

TURKEY MULLEIN

Eremocarpus
  setigerus   Euphorbiaceae

OTHER NAMES: Dove Weed

APPEARANCE: This is a neat, silvery, little annual, with shallow roots and dense, coarse, hairy foliage. The odd-scented leaves are rounded and strongly veined, with a slight concave upward curl; leaves pile one on the other in mad abandon to photosynthesize while the moisture holds out and form as many little terminal yellow flowers and round seeds as possible. When gathering or handling the plant, you will notice how the leaf hairs are almost (but not quite) stinging, and you may itch a bit afterwards. They line the roads, mile after mile, in most of California.

HABITAT: Roadsides (mile after mile . . .), from the western and central Mojave Desert to the coast and up to southern Washington. The plants growing in the California deserts are much higher in substances. I don't know where they grew before we built roads.

CONSTITUENTS: Eremone, inautriwaic acid (both diterpenes), b-pinene, myrcene, trans-ethyl-cinnamate.

COLLECTING AND PREPARATION: The whole plant in flower, late spring or early summer, tinctured fresh, 1:2, in vinegar, or dried and chopped for later use.

MEDICINAL USES: The fresh plant vinegar makes an excellent counterirritant for chest and intercostal-muscle soreness caused by physical exertion or heavy coughing. It helps the pain of dry-membrane pneumonitis and pleurisy, as well as mild emphysema and the "smog lungs" that some children and adults get from the humid air pollution in some of the Southwest.

The fresh plant tincture applied to the forehead and scalp can help some sick headaches and hangovers. The Pomo and Kawaiitsu of California used it as a well-strained tea (don't want to swallow those itchy hairs) for stress palpitations, hot and cold shaking chills from viral fevers, and for bleeding diarrhea. A tea of the dried plant added to bath water will stimulate sweating in early, dry fevers, especially in children. Allow half of their body to be out of the bath water in order to allow the sweat to lower their temperature; keep them warm afterwards.

Drinking the tea for chills seems to work better when it is boiled as a decoction, evaporating off some of the volatile oils and leaving the nonevaporative muscle-relaxant constituents still in the tea.

I personally don't know how much tea it takes to cause any side effects; the family it comes from, the Euphorbias, has many potentially toxic plants in it. I have drunk fair amounts of the tea at one time to see what would happen, but I just sweated so much that I needed a blotter. This is called organoleptic testing, and it is probably why herbalists, homeopaths, and natural-products chemists have trouble getting medical insurance.

All traditional uses point to the use of Turkey Mullein for short periods of time, during periods of acute symptoms, not for any extended use. Sounds reasonable.

OTHER USES: This is the main fish poison used by California Indians. It was once assumed that those effects were mechanical, the hairs clogging up their gills and making them float sideways. Actually, the diterpenes seem to be the main cause for the plant's effects on fish. The one time I tried to check it out and find out how well it worked, I could only find three little fish in all of Topanga Creek, and they didn't look too well to begin with. Anyway, how do you get fish to stay still while you dump a bunch of silage on them?

VERVAIN

Verbena ambrosifolia, V. wrightii, V. bipinnatifida, and
others   Verbenaceae

OTHER NAMES: Blue Vervain, Desert Verbena, Dormilón, Moradilla, Verbena

APPEARANCE: These are low, hairy, somewhat spreading plants. The stems are usually less than a foot in length, crowding out from the stem to form showy purple puffs or scraggly, over-the-hill, laying down, bald in the center desert shrublets . . . all dependent on how much water has been around. At their best (and most pickable), these plants have opposite hairy leaves, deeply cleft, with terminal flowers forming flattened puffs of magenta, purple, or pink, blooming from early spring to fall. Small plants, but with recent rain they may form conspicuous showy displays along our desert roads and middle-altitude hillsides.

HABITAT: Found at all altitudes, from Barstow, California north to Salt Lake, east to the Panhandle, south to central Texas, usually from 2,000 to 5,000 feet but higher in central New Mexico, lower in southern Arizona. Abundant.

CONSTITUENTS: Stachyose, verbenaloside (cornine), verbenalin (splitting into verbenalol and hastatoside), the iridoids aucubin and artemetin, B-sitosterol, ursolic acid.

COLLECTING: The flowering stems, dried Method A after lightly washing and shaking.

STABILITY: At least 1 year.

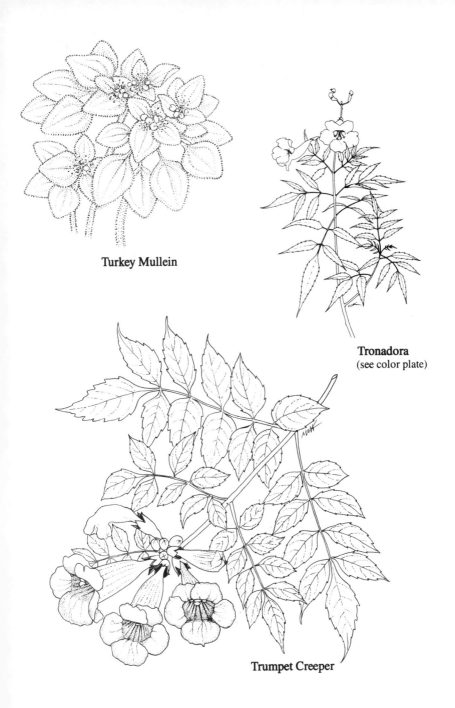

Turkey Mullein

Tronadora
(see color plate)

Trumpet Creeper

127

PREPARATION: Standard infusion, 1–3 ounces; Method B tincture, 1:5, 50% alcohol, ¼–½ teaspoon.

MEDICINAL USES: Vervain is broadly active medicinally, serving as a sedative, diaphoretic, diuretic, bitter tonic, antispasmodic, and mild coagulant. It is one of our best palliatives for the onset of a virus cold, particularly with upper respiratory inflammation. It will promote sweating, relax and soothe, allay feverishness, settle the stomach, and overall produce a feeling of relaxed well-being. It is especially useful for children who become fidgety and cranky when first feeling ill, run around in a droopy, hyperactive fashion, but . . . will . . . not . . . sleep. The dose of a simple infusion for them is one-half to one tea-spoon of the herb in ½ cup of water, or 20–30 drops of the tincture in a little warm water. For adults use the larger doses mentioned earlier.

Vervain is a useful bitter tonic for nervous stomachs, inhibiting the sympathetic nerve endings that themselves inhibit, under stress, the normal digestive secretions and muscle contractions of the stomach. The herb has shown moderate stimulus in humans to intestinal secretions, salivary output, and milk formation, slightly stimulating the force of cardiac output and blood-vessel tone.

It is sometimes useful to sip the tea in difficult births, flavored with some pleasant mint and a little honey, since it mildly stimulates the uterus tone and strength of contractions, is mildly vasoconstricting, and will help limit postpartum bleeding as well as placenta separation.

A nice plant.

WILD LETTUCE

Lactuca serriola   Compositae

OTHER NAMES: *L. scariola,* Prickly Lettuce

APPEARANCE: This is a scruffy roadside weed with two- or three-inch variously lobed or pinnatifid leaves clasping the stem and twisting sideways to become perpendicular with the ground. The central vein of the leaves is thick and riblike, prickly on one side and distinctly lighter than the dark blue-green of the rest of the plant. The height is generally two to four feet, forming one stem until the annual starts to flower into insignificant, yellow Dandelion-like flowers; then the single leaving stem divides into many small, thin floral stems. The whole plant has milky sap, like Dandelion, Sow Thistle, and Chicory (all related). Pretty distinctive plant for one so unaesthetic.

HABITAT: Likely to be anywhere in waste areas below 7,500 feet, although a stronger medicine in our lower, drier desert climate. It grows in all the states in our area and is most abundant from 4,000 to 6,000 feet, along roadsides and in gravelly junk places. An annual weed that forms large, dispersed stands in one county and is virtually nonexistent in the next.

CONSTITUENTS: Lactucin, lactucic acid and other odds and ends, none of which account for its mild analgesic effects.

COLLECTING: The whole plant in early blooming, dried, Method A. For collecting the dried latex ("Lactucarium"), find a nice stand of the plants in emerging or full bloom, bring a chair, some shade from the sun, a picnic lunch in a cooler, a good book, a sharp scalpel, straight razor or microtome knife, and a round Tupperware bowl. Slice all the flowers from every stem on at least fifty plants, let the milky sap "set" a bit until slightly browned (about a ten-page chapter in the book), scrape off the latex from each cut into the bowl and slice each one about ½ inch below the first slice, let the sap harden a bit in the air (another chapter in the book), scrape into the bowl, slice each and every stem again . . . on . . . and . . . on. Figure a yield of about 2–3 grams from a four-foot plant after it has been systematically bled until reduced to a one-foot-tall, headless pygmy and the latex dried down in a shady but breezy place until it resembles a mixture of butterscotch pudding, dried library paste, and tubercular phlegm. This is called "the harvesting of the Lactucarium" and is so boring and tiresome that even *I*, a little bit fanatic about botanical medicines, have only managed this harvest twice in my two decades of somewhat manic wildcrafting. Take this questionable batch of dried rubber cement and dissolve it in 80% alcohol at a ratio of 1:5. Doses for the dried herb, as a standard infusion, are from 2–8 ounces as needed; of the tincture latex, ½–1 teaspoon up to four times a day.

MEDICINAL USES: Wild Lettuce tea and the tincture are feeble analgesic narcotics and as such are very useful where there is a need for a mild opiate in children who cannot sleep from constant coughing or from intestinal cramps. Although self-limiting in its strength (it never seems to rise above the epithet "feeble"), it depresses spinal-chord referral of pain and sedates the sympathetic ganglia of the thoracic parts of the autonomics. It is useful for adults as well, especially when used in synergy; Valerian for the wakefulness, Passion Flower or Silk Tassel if there are cramps, Black Cohosh for muscular or joint pain.

OTHER USES: The dried latex, smoked like opium, brings to mind, by taste at least, the reason for Lactucarium once having been referred to as Lettuce Opium. Other similarities are probably placebo only.

WILD OATS

Avena fatua   Graminaceae

APPEARANCE: This is an easy plant to identify if you are at all familiar with the cultivated variety. Wild Oat is an annual grass from one and one-half to three feet high. The leaf blades are long, broad, and rough; the sheaths are smooth or slightly hairy. The flowers have a typical oat open inflorescence, the spikelets are two- or three-flowered. The ripening seeds hang out and down, covered in bristles, with a twisted awn. If the seed is smooth and the awn straight you have come across some cultivated oats (*Avena sativa*) and it's used the same way. If the seeds are small, brown, and bristly and look like a little cockroach (with two feelers *and* an awn) it is Slender Oats (*A. barbata*), which doesn't make good medicine; so forget it.

Vervain

Wild Lettuce

Wild Oats

130

HABITAT: Everywhere there is dirt, moisture, and a spring season; it especially likes sloping vacant lots and semideveloped rural suburbs with lots of recently bulldozed disturbed earth.

CONSTITUENTS: Ergothioneine, histamine, hordenine, trigonelline, avenin, avenacines A and B.

COLLECTING: You want the semimatured grain, when in "milk"; by the time the seeds in the middle of the stalk squirt white, the whole seed head can be stripped from the stalk and used. This usually happens in May, but sometimes earlier or later.

PREPARATION: The unripe-seed tincture, Method A. Since this is the only form in which the medicine works, if you live in California and don't have any 95% alcohol available to you, use 50% vodka (100 proof). Better inferior than nothing, right?

MEDICINAL USES: Before you say to me (which I can't hear across the reader-writer chasm), "Tincture of green Wild Oats . . . are you out of your mind, Moore?" let me quote to you from the 1929 *Handbook of Pharmacy and Therapeutics,* published by Eli Lilly and Company, regarding how to use their preparation of fresh Avena: "Tonic, laxative, and nerve stimulant. Used in chorea, epilepsy, nervous exhaustion, and in the treatment of habitual narcotism." See! The value of the therapeutic hasn't changed, just current medical practice. It is a beautifully effective medicine in the depressive states that "yangy" people get, those headstrong and bodystrong people, burned-out emotional crispy critters with not much life experience in handling their depressions. Maybe a death in the family, maybe a breakup of marriage, whatever it is, the depression caught them by surprise, and they don't know how to handle it. Wild Oats, ¼ teaspoon as needed. It also helps cocaine and amphetamine burnout and those adrenalin-dominated persons that get stressed out, constipated, and grumpy. Or picture the parents, busily preparing for the "Big Vacation" . . . wondering by the first evening on the road why they even bothered to procreate. (Someone in the backseat is yelling for the 432nd time, "Are we there yet, Mommydaddy?") Wild Oats.

YERBA DE ALONSO GARCIA

Dalea formosa   Leguminosae
OTHER NAMES: Perosela formosa, Feather Dalea

APPEARANCE: This is a little, scraggly shrub, generally two or three feet tall, formless and arbitrary in shape, sometimes barely more than a few senescent twigs. The leaves are small and feathery, grey-green and pinnate or semicompound. The branches are brittle-woody and sort of spiny, small, and slender.

There is a fragile, ancient visage about them, as if they were born old and are barely hanging in there. The flowers emerge at the ends of the twigs and short branches, up to half a dozen in a bunch, and are nestled in long, silky-haired calices. They are a striking midnight magenta, resembling sweetpeas over-dressed for a graduation dance. When you have gathered a bunch of the flowering and leafing branches you notice its secret . . . the sweet delicious smell, like honeysuckle and jasmine.

HABITAT: Widespread along grassy arroyos, roadsides, and gravelly slopes, from nearly as far west as Kingman, Arizona, east to the Panhandle, north to Santa Fe, and all points south. It is especially abundant in the eastern Gila and lower Rio Grande drainages of New Mexico. From 3,000 to 6,500 feet. And no, I have no idea who Alonso Garcia was.

COLLECTING: The flowering branches in April and May, dried in a paper bag.

PREPARATION: A simple tea, brewed for the flavor.

MEDICINAL USES: I have included this little lovely in the book because it makes one of our best tasting herb teas, sweet, delicate, and a little tart. You can brew it fresh from the plant, filling a jar with the flowering branches, covering them with water, setting it in the sun for a couple of hours, and drinking it. The fragrance stays for months and is heartening to the soul drunk hot and scented on a cold winter night.

Pueblo Indians and the Apaches used it as a treatment for growing pains and aching bones. The Hopis use it for influenza and virus infections, considering it a "cold" herb for hot conditions. New Mexican Spanish will make a strong bath with the branches and bathe in it for a couple of hours to relieve arthritic pains. I don't know how well it *really* works, but I have taken such a bath, and the scent must have some subtle beneficial effect to the cerebellum, if nothing else.

YERBA MANSA

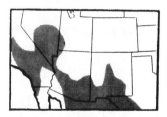

Anemopsis californica   Saururaceae
OTHER NAMES: Yerba del Manso, Manso,
Lizard Tail, Swamp Root

APPEARANCE: Yerba Mansa is a low-growing plant always found in stands. It has many basal leaves from three to six inches long, rounded, and succulent, often with a reddish-silvery bloom, especially on the stems and leaf edges. The flowers form conical white spikes with strikingly white bracts around the base, overall resembling a coneflower or Anemone. The whole plant turns brick-red in the fall, with foot-tall seed-stalks mingled in masses of long, brick-red basal leaves. The dead foliage is the same shade as dead Yellow Dock. The leaf stems root freely at the nodes and often form interconnected colonies like strawberries. All parts of the plant are strongly aromatic, especially the fleshy root, a cross between Camphor and Eucalyptus. New leaves start growing from the root crowns in early spring, flowering takes place in May and June, the plants start turning brick-red in September.

HABITAT: Yerba Mansa *loves* swamps, oozy glurches, and the alkaline mires that even cattails won't grow in. Once the colony is established, however, the soil chemistry and organisms are altered and cattails *start* to grow, since a few decades of dead leaves and root oils improve, acidify, and aerate the soil, inhibit microorganisms, and broaden the ecology. They are found in moderate abundance in the central valley of California (don't gather there; too much pollution from agribusiness chemicals), Imperial Valley (same problem), in saline swamps in southern California, the waterways of the upper Colorado River, the middle elevations of Nevada, the Salt River and Gila River drainages, all along the Rio Grande basin, from Española, New Mexico, nearly to the Gulf Coast, and moist marshes in Sonora and Chihuahua.

CONSTITUENTS: Methyleugenol and related compounds, esdragole, thymol methylether, linalool, p-cymene, asarinin, and other aromatics.

COLLECTING: The roots, dug at any time of the year, although they are strongest when the leaves have died back to brown in the fall and winter. Use a stout shovel, since the clay silt it prefers is intractable. If you find it growing in nice, sane sand, consider yourself fortunate indeed . . . and write to me about the location! Wash the roots well, wilt them in flats for several weeks, then cut into small sections and continue to dry. If you slice them fresh they are apt to mold; if you let them dry intact (it can take six months), they can turn dark, corky, and lignaceous . . . totally useless.

STABILITY: Small, dried-root sections are usually good for at least eighteen months . . . longer if the characteristic smell is still strong.

PREPARATION: Fresh root tincture, Method A; dry root tincture, Method B, 1:5, 60% alcohol; recently ground dry-roots for dusting, antiseptic washes, and tea. Capsules are fine, but the aromatics start to go after a month unless bagged and frozen.

MEDICINAL USES: *Everybody* who has lived where Yerba Mansa grows has used it as a medicine. It was used in standard practice medicine with some frequency up until 1920 or so, especially amongst Eclectic and Homeopathic physicians. Although not related botanically or chemically to Golden Seal (*Hydrastis canadensis*), much of what it does is similar, particularly in regard to subacute congestion in the mucous membranes.

When inflammation resulting from irritation, injury, or infection continues past a certain point, the engorged capillaries lose some of their cohesiveness. This breaks down the quality of the gelatinous starches (mucopolysaccharides) that hold tissue cells into compact masses, thwarting the healing that the initial irritation made necessary. The tissues go from the "hot" of acute (repairing) inflammation to the "cold" of subacute (nonrepairing) congestion. Thus, to the New Mexican Spanish and the Pueblo Indians of the Southwest, Yerba Mansa (hot) is used for boggy, poorly healing infections (cold). This dualist approach is also found in Chinese, Japanese, Ayurvedic, and Unani medicines. So, Yerba Mansa is used for slowly healing boggy conditions of the mouth, intestinal and urinary tracts, and lungs. It is astringent to the connective tissues that form the membrane structure, but it stimulates better fluid transport, helping to remove the exudates that prevent repair of the irritation that began the whole mess.

Mouth, gum, and throat sores are helped by the herb, as are ulcers of the stomach and duodenum. The feeling of digestive hangover after intestinal infections responds well to the tincture, tea, or capsules. Use ¼ teaspoon of either tincture in water, a Standard Infusion, 2–3 ounces, or 2 #00 capsules, 2–3 times a day. For boggy membranes in the intestinal tract, combine with Echinacea; for Giardia or amoeba infections, start with Chaparro Amargosa, finish with Yerba Mansa; for cystitis or urethritis, start with Cypress, Grindelia, or Prickly Pear (acid urine), Manzanita (alkaline urine), move to Yerba Mansa, finish off with some soothing Hollyhock or Malva.

As a diuretic, Yerba Mansa stimulates the excretion of nitrogenous acids, especially uric acid, which can aid many types of joint problems. It is also substantially aspirin-like in its anti-inflammatory effects, and, with both these things happening, it's no wonder that it is such a widely used folk medicine in the Southwest for arthritis and the like. Combine with Dandelion Root or Shepherd's Purse for the increased uric-acid secretions, with Yucca, Agave, or Puncture Vine for an internal anti-inflammatory, and with Matarique, Camphor Weed, or Escoba de la Vibora for an external anti-inflammatory.

It is antibacterial and antifungal; so it affords a fine external first aid or dressing for abrasions or contusions. A sitz bath for bartholin gland cysts and perianal fissures or boils usually brings quick healing; use 1 teaspoon of the tincture per quart of water, or a 1:64 decoction of the powdered root. The powdered root is an impeccable dust when mixed with four parts of a soothing starch for diaper rash and chafing. The leaves, although much feebler and chemically simpler, make a fine bath for general pain of the muscles and joints; combine with Jimson Weed, Tobacco, or Pineapple Weed.

A water percolation (1:10), with 20% glycerine and 10% alcohol added when finished, is an excellent nasal spray for hay fever, lingering head cold, or the results of cocaine or snuff abuse.

Finally, used by itself (powdered root) or combined with Cypress and Chaparral, is is excellent for athlete's foot.

YUCCA

Yucca spp.   Liliaceae

OTHER NAMES: Amole, Spanish Bayonet, Joshua Tree, Datil, Spanish Dagger

APPEARANCE: This common and ubiquitous plant is distinct and easy to recognize. The various Yuccas have numerous spiny-tipped elongated leaves that rise in a cluster from a central stem, usually from ground level but in several species from one or more trunks. The long leaves are constantly shedding along the margins but are otherwise only armored at the leaf tips. Anyone who has backed into the aptly named Spanish Bayonet will vouch for the simple efficiency of the barbed tips, however. Often confused with Agave and Beargrass, but the former has broader, thick, frequently spined leaves and a tall, obliquely branched flower stalk and the latter has much thinner, grasslike leaves that are often seen lying partially on the ground. Yucca has either a single two- to five-foot flower stalk, upright and conspicuous, or is branched like the California Joshua Tree with one flower stalk for each arm. The flowers are lilylike, either large and cream colored with brown flecks or smaller and yellow green. They usually open at night, closing downward in the daylight. Some of the Yucca fruit is large and succulent, particularly *Y. baccata*, but the taste tends to be mealy-bland.

HABITAT: Found most frequently in stands, often covering valleys and dry mountain slopes, it is most common in the high desert, extending up into the Juniper/Piñon belt and higher still on drier slopes in the Ponderosa belt. From nearly sea level in California to 8,000 feet on the dry sides of mountains in Utah, Arizona, New Mexico, and Colorado, throughout the West, sparse farther north.

COLLECTING: The root at any time of the year. Should be split lengthwise (Method B) before drying. If for medicinal use the bark may be removed; if for washing or rinsing the hair, the bark should be left on. Use only after drying.

MEDICINAL USES: At one time considered a potential source of phytosterols, a family of plant substances used in the manufacturing of steroidal hormones, its present use is as a sudsing agent in the cosmetic and soap industries and as a home remedy for arthritic pain. Recent clinical studies have shown it to be of some use in the treatment of joint inflammations but the function is not understood. One-quarter ounce of the inner root should be boiled in a pint of water for fifteen minutes and drunk in three or four doses during the day. Arthritis being such an idiosyncratic disorder, no single treatment will help more than a percentage of people, but if Yucca tea is effective, it can relieve pain for several days afterwards. If a strong laxative effect persists, especially if accompanied by intestinal cramping, decrease the amount next time. If there are no side effects, the quantity can be increased to one-half ounce a day. Long-term daily use can slow absorption in the small intestine of fat-soluble vitamins. The tea has some value for urethra or prostate inflammations.

OTHER USES: Added to shampoo or used by itself for washing dry hair. One-half to one cup of the chopped fresh or dried root is boiled in one and one-half cups of water until suds form.

CULTIVATION: From roots dug in the late fall and replanted in well-drained sandy soil.

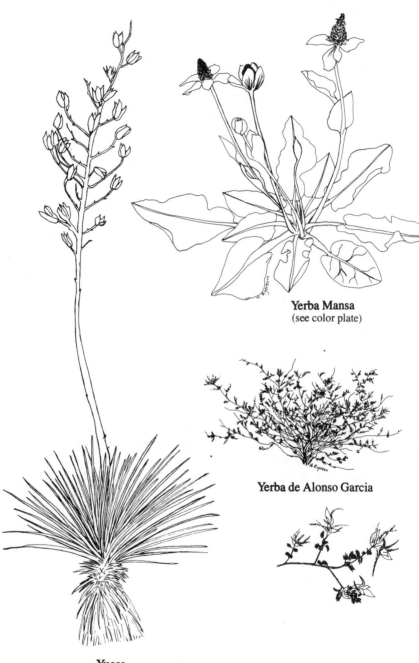

Yerba Mansa
(see color plate)

Yerba de Alonso Garcia

Yucca
(see color plate)

FORMULAS

The specific herb that describes the formula is listed in upper case.

1# HEMORRHOID OINTMENT

AÑIL DEL MUERTO	2 parts by weight
Echinacea Flowers OR	
Yerba Mansa	2 parts
Tobacco OR Jimson Weed	1 part

Powder together, make into a salve, Method A.

#2 LIPID NORMALIZER

CALIFORNIA MUGWORT	2 oz. cold infusion (approx 2 g.)
Algerita	40 drops tincture
Milk Thistle	40 drops tincture

Combine the tinctures with the tea and drink before retiring.

#3 CHRONIC BLEEDING FORMULA

CANADIAN FLEABANE	2 parts by weight
Milk Thistle	1 part
Prickly Pear Flowers	1 part
Shepherd's Purse	2 parts

Make a standard cold infusion, take 2–3 oz. up to three times a day.

#4 SUNBURN TREATMENT

CAÑAIGRE—fresh root, grated on the burned area, allowed to dry, followed by fillets of Prickly Pear pads or the freshly pressed sap. Repeat as needed.

#5 FLYPAPER

CHAPARRO AMARGOSA	4 tabl., finely chopped
Honey	1 cup
water	2 tabl.

Heat slowly for 1 hour at a very slow boil. Strain, dissolve 1 oz. of beeswax in the hot mixture, cool until uncomfortably but bearably hot, spread evenly with a spatula on freezer paper or lightly oiled paper. Cut the flypaper carefully with a razor blade or utility knife and hang.

#6 MASTITIS FORMULA

COTTON ROOT	3 part
Echinacea	2 parts
Inmortal Root (see *MPOMW*)	1 part

Tincture together (1:5, 60% alcohol) or mix the separate tinctures. Use 1–2 teaspoons in ½ cup of hot water, sip slowly, and repeat in several hours if needed.

#7 URINARY DISINFECTANT
CYPRESS
Manzanita
Yerba Mansa
Shepherd's Purse OR Hollyhock
  Root
Equal parts by weight, brewed as a cold standard infusion (or hot if necessary for quicker use); drink up to a quart of tea a day.

#8 IMMUNOSTIMULANT

Echinacea Tincture	60 drops
INDIAN ROOT	30 drops
Hollyhock Root (cold standard infusion)	4 oz.

During active cold, flu, or other acute infections, take this combination every three or four hours. As the infection breaks, replace every other dose with good Alfalfa tea, then recuperate totally with the Alfalfa and stop the other three herbs.

#9 EMOLLIENT POULTICE ("Species Emollientes, N.F.")
Hollyhock Leaves
MALLOW LEAVES
Sweet Clover (see *MPOMW*)
Chamomile (or Pineapple Weed)
Flaxseed
Use equal parts by weight, powder together and refrigerate or freeze any unneeded portion of the mixed powder to prevent loss of aromatics by evaporation. To make the poultice, add hot water equal to four times the volume of powdered herbs, allow to set slightly, and apply the hot mash directly to the skin.

#10 INTESTINAL DISINFECTANT
MESQUITE LEAVES or PODS
Desert Willor OR Tronadora OR Trumpet Creeper
Chaparro Amargosa
Echinacea
Silk Tassel Leaves.
Use equal parts (by weight) for a cold standard infusion, take in 4-oz. doses, up to a quart a day of the tea.

#11 HYPERTENSION FORMULA (in the strong, sthenic, middle-aged person)

PASSION FLOWER	¼ tsp. tincture OR ⅛ oz. herb
Hawthorn Berries	¼ tsp. tincture OR 1 tsp. berries
L-tryptophane, 250 mgs.	

Take tincture or tea with the amino acid supplement before retiring... it starts to help after about a week of the regimen.

#12 MOUTHWASH (for gums, mouth, or dentures sores)

RATANY	2 parts by weight
Elephant Tree Gum (or Myrrh Gum)	1 part by weight
Yerba Mansa	2 parts by weight
Sumach Leaves	1 part by weight

Tincture together in 75% alcohol (or Myer's Rum, 150 proof), 1:4, adding 15% glycerine to the finished tincture.

#13 DIARRHEA FORMULA

RATANY
Echinacea
Silk Tassel Leaves
Trumpet Creeper OR Desert Willow
  OR Tronadora

Use equal parts by weight to make a cold standard infusion, and drink ⅓ cup every several hours, up to 3 cups a day. Best for food- or water-derived infections.

#14 NIGHT SWEAT FORMULA

SOAPBERRY LEAVES	Cold infusion, 2–3 oz. (2 gm. herb)
Indian Root Tincture	30–40 drops
Silk Tassel Leaf Tincture	30 drops
Echinacea Root Tincture	30 drops

Take the tinctures in the tea once late afternoon, once before retiring.

#15 ALTERATIVE SYRUP

STILLINGIA	4 parts by weight
Echinacea Root or Flowers	4 parts by weight
Yellow Dock (see *MPOMW*)	3 parts
Algerita (or Oregon Grape, see *MPOMW*)	2 parts
Yerba Mansa	2 parts
Milk Thistle	2 parts

Tincture the herbs, ground together, to 1:4 in 50% alcohol, add

Glycerin	2 parts by volume
Honey	2 parts by volume

Use the syrup in one- or two-teaspoon doses.

GLOSSARY

ABORTIFACIENT: Any substance used, intentionally or otherwise, that may induce a miscarriage.

ACETYLCHOLINE: A choline ester found in many tissues of the body, most specifically the nervous system, where it facilitates the transmission of nerve impulses between nerve cells (at the synapses), and between nerve endings and most muscles and glands.

ACNE: A chronic skin problem, usually in the head and torso, manifesting itself through eruptive inflammations of the sebaceous glands and hair follicles. It may be caused by steroid hormone imbalances, food allergies, dietary deficiencies (particularly vitamin A), indigestion, stress, and faulty metabolization of blood lipids by the liver.

ACUTE: A type of disease having a sudden onset, severe symptoms, and a generally short duration (e.g., a head cold). The opposite of chronic.

ADENOVIRUS: A family of viruses, simple and nonenveloped, that cause acute respiratory problems in humans, as well as in other mammals and birds. Unlike some other viruses, the adenoviruses induce a complete antibody/antigen immunologic response, and poor resistance to them is often a symptom of impaired immunity or physiologic stress.

ADRENAL CORTEX: The outer covering of the adrenal glands, the two triangular endocrine glands that rest atop the kidneys. Formed in fetal development from the same tissue that becomes the gonads, the adrenal cortex secretes a number of steroidal hormones that regulate carbohydrate use, salt balance, reproductive functions, anabolism, catabolism, and inflammation. In Chinese medicine, these functions relate to kidney yang. Inside the cortex is the adrenal medulla, which is the source of epinephrine (adrenalin) and relates to kidney yin.

ADRENERGIC: Those responses, sometimes called "flight or fight," that are triggered by epinephrine (adrenalin). These responses are the result of the stimulation of specialized nerves that respond to epinephrine instead of the local neurohormone acetylcholine and can therefore be stimulated *en masse* by fear or anger. The more frequently a person triggers this response, the more dominant become the adrenergic nerves; anti-andrenergic drugs (beta-blockers, as an example) are used in controlling the elevated cardio-vascular functions induced by chronic adrenergic responses.

ADRENOCORTICAL: Of, and pertaining to, the adrenal cortex.

ALKALOID: One of a varied family of alkaline, nitrogen-bearing substances, usually plant-derived, reacting with acids to form salts. Usually intensely bitter, alkaloids form a body of substances widely used in drug and herbal therapy. They generally have a toxic potential. (Examples: caffeine, morphine, berberine.) The term is more pharmaceutical and medical than chemical since alkaloids come from a variety of otherwise unrelated organic compounds.

ALLOPATHY: A term loosely applied by other therapies to the form of medicine practiced by physicians. More appropriately, the use of drugs or other means to antidote a disease or symptoms, e.g. using aspirin to lower a fever. Since medical doctors can use many approaches to dealing with a disease, the term only applies to a certain part of their therapeutics. It is generally a perjorative term applied to standard practice medicine. The opposite of homeopathy.

ALPINE: Those plants found predominantly at and above the timberline.

ALTERATIVE: A term applied in naturopathic, eclectic, and Thomsonian medicines to those plants that stimulate changes of a defensive nature in metabolism and tissue function in the presence of chronic or acute disease. The whole concept of alteratives is based on the premise that disease symptoms in an otherwise healthy individual are actually the external signs of activated internal defenses and, as such, should be stimulated and not suppressed. For example, Marsh Fleabane acts as an alterative when used to stimulate sweating in a fevered state, although used otherwise, Marsh Fleabane will not induce either the sweating or the fever. Its function as an alterative exists only when the body itself has begun a fever as a defense mechanism; without the change of metabolism called a fever, the drinking of Marsh Fleabane tea serves only to be a mild diuretic. Alterative is sometimes inaccurately used as a synonym for "blood purifier," a term better applied to the liver and spleen and not some dried plant.

ALTERNATE: A botanical term denoting an arrangement of plant organs (usually leaves) that alternate along opposite sides of a stem, as distinct from paired or whorled.

AMINO ACID: A large group of organic, nitrogen-bearing compounds. They form the building blocks from which life assembles, on the cellular level, the various complex proteins. In humans, unused amino acids are excreted primarily in the urine and sweat as urea.

ANABOLIC: Promoting anabolism. Specifically, an agent or function that stimulates the organization into more complex substances from less complex substances. For example, making proteins from amino acids is anabolic; making glycogen from glucose or triglycerides from fatty acids and glycerol is also anabolic. Anabolic steroids are those substances that, endogenous or exogenous, induce increased body size. The opposite of catabolic. The two extremes, anabolic (growing) and catabolic (shrinking), form metabolic homeostasis.

ANALGESIC: A substance that relieves pain. (Examples: aspirin, Camphor Weed.)

ANESTHETIC: A substance that decreases sensitivity to pain. (Examples: nitrous oxide, Oil of Cloves.)

ANODYNE: A substance that relieves pain, usually with accompanied sedation. (Examples: morphine, Prickly Poppy.)

ANORECTIC (ANOREXIC): A substance or condition causing a decrease in appetite for food.

ANTABUSE: The most common trade name for disulfiram, a drug used as an alcohol deterrent in the treatment of chronic alcoholics. It is also a moderate immunosuppressant.

ANTIMICROBIAL: An agent that prevents the growth of microbes.

ANTIOXIDANT: Specifically, a substance that prevents oxidation or slows a redox reaction. More generally, a substance that slows the formation of lipid peroxides and other free-radical oxygen forms, preventing the rancidity of oils or blocking damages from peroxides to the mitochondria of cells or cell membranes. (Examples: vitamin E, Chaparral, NDGA.)

ANTISEPTIC: A substance that will prevent or retard the growth of micro-organisms.

ANTISPASMODIC: A substance that will relieve or prevent spasms, usually of the smooth muscles of the intestinal tract or uterus. (Examples: barbiturates, Silk Tassel.)

ARRYTHMIA: An irregularity of rhythm, usually in reference to the heart.

ARTHRITIS: Inflammation of one or more joints from any number of causes, usually accompanied by some alteration of joint structure.

ARTERIOSCLEROSIS: The condition of blood vessels that have thickened, hardened, and lost their elasticity; "hardening of the arteries." Aging and the formation of fatty plaques within or directly beneath the inner lining of the blood vessels are the most common causes.

ASTHMA: A chronic disease characterized by paroxysms of the bronchi, labored breathing, and wheezing. There are sometimes allergic causes, sometimes emotional causes, usually a mixture of factors. The bronchial bore is normally smaller and contracted at rest, when it is under parasympathetic or cholinergic dominance (yin); the bore increases during physical activity, when it is under sympathetic or adrenergic control (yang). Uncontrolled spasms of contraction and expansion are the usual form that asthma takes, and the most common medical approach is to force the bore open with adrenergic-mimicking drugs.

ASTRINGENT: An agent that causes the constriction of tissues and is used to stop bleeding, secretion, and the like. (Examples: a styptic pencil, cold temperatures, Ratany.)

ATHEROSCLEROSIS: Arteriosclerosis caused by the formation of fatty plaques (atheromas) within or directly underneath the inner lining of blood vessels. Faulty synthesis of fats and lipoproteins by the liver is the usual cause.

AUTO-IMMUNITY: An abnormal condition in which the body produces antibodies or has an abnormal white blood cell response to itself or certain of its proteins, antibodies, or immunoglobulins. Diseases characterized by auto-immunity include rheumatoid arthritis, myasthenia gravis, multiple sclerosis, membranoproliferative glomerulonephritis, hemolytic anemia, and dermatomyositis. Although inherited factors (such as tissue type) allow for the possibility of forming an auto-immune response, most overt forms seem to be triggered by metabolic stress, certain virus infections, environmental stress, and trauma.

AXILLARY: Borne in the axil, that is, between the leaf and the stem. In anatomy, refers to the armpit.

BALSAMIC: Soft or hard plant or tree resins composed of aromatic acids and oils. These are typically used as stimulating dressings and aromatic expectorants and diuretics. This term is also applied loosely to many plants that may not exude resins but have a soothing, "pitchy" scent. (Examples: Elephant Tree, Encelia, Marsh Fleabane.)

BARBER'S ITCH: Inflammation and infection of the facial hair follicles, either by staphylococcus (folliculitis barbae) or by a skin fungus (tinea barbae).

BIENNIAL: A plant that forms (usually) a basal rosette of leaves the first year of growth and a flowering stalk the second year. Most true biennials die the second fall, but there are many variations in cycle under different growing conditions. Many plants that are normally annual may, in the warmer deserts, flourish for two or more years; these do not usually form the typical biennial pattern of growth.

BILE: A digestive and excretory secretion from the liver. Bile stored in the gall bladder is more concentrated and may be green or brown in color, whereas bile secreted directly from the liver is watery and straw-colored, like urine. Bile excretion is stimulated by the presence in the duodenum of fats and when combined with alkali from the pancreas serves as a detergent to emulsify them and make them more water soluble. Bile also serves to stimulate peristalsis in the large intestine, colors the feces brown, and is the medium for excreting hemoglobin waste products and some surplus cholesterols.

BILIARY: Pertaining to the bile secretions.

BILIOUSNESS: A symptom-picture resulting from a short-term disordered liver, with constipation, frontal headache, spots in front of the eyes, poor appetite, and nausea or vomiting. Usual causes are heavy alcohol consumption, poor ventilation when working with solvents, heavy binging with fatty foods, or moderate consumption of rancid fats. The term is genially archaic in medicine; people who are bilious are seldom genial, however.

BILIRUBIN: An orange-yellow pigment that is secreted in bile as the final unusable end product of hemoglobin recycling. It contributes to the brown color of feces and, in the case of bile duct or liver malfunctions, backs up into the blood to cause jaundice.

BIOFLAVONOIDS: A group of substances formerly called vitamin P, usually associated with ascorbic acid in natural sources. The "bio-" is redundant; they are usually called flavonoids by chemists. They are brightly colored pigments and, in nutrition, are usually referred to as lemon or citrus bioflavonoids . . . or simply as the "complex" part of vitamin C complex. They are widely used to increase the rigidity and strength of capillary walls, in conditions ranging from stomach ulcers to whiskey nose; they have only marginal use in American medicine. (Examples: rutin, hesperidin, citrin, Prickly Pear Flower.)

BITTER TONIC: A bitter-tasting substance or formula used to increase a deficient appetite, improve the acidity of stomach secretions, and slightly speed up the orderly emptying of the stomach. A good bitter tonic should have little, if any, systemic effects other than on oral and stomach functions and secretions. A bitter tonic will have little, if any, effect on individuals with normal digestive functions.

BLOOD PURIFIER: A loose and simplistic name for those herbs which seem to speed up the detoxifying and excretion of waste products in the bloodstream, particularly when there are resultant skin eruptions. The apparent value in so-called blood purifiers may come from a stimulation of liver functions or bile secretions as well as through intestinal stimulation when the eruptions are digestion-related. A vague but pragmatic term.

BRADYCARDIA: Slow heart action, most frequently defined as a pulse below sixty beats a minute (seventy per minute in children).

BRONCHIAL DILATION: Relaxing and opening of the upper parts of the lungs to improve inspiration and relax constricting spasms.

BRONCHIECTASIS: An abnormal dilation of the cartilage-ringed branches of the trachia, the bronchi. This term may also include dilation of the smaller branches, or bronchiols (a condition termed bronchiolectasis). Although sometimes congenital, it can be an acquired condition resulting from obstructions, bronchial infections, chronic bronchitis, and other extended bronchial disorders. These dilations are usually mucus- or pus-filled and may become fibrous with time.

BRONCHITIS: Inflammation of the mucous membranes that line the bronchi.

BRONCHORRHEA: Excessive mucus secretion by the bronchial linings; the lung equivalent of a runny nose.

BURSITIS: Inflammation of a bursa, the lubricating sac that reduces friction between tendons and ligaments or tendons and bones. The most common types are bursitis of the shoulder, elbow, knee, and big toe (a bunion).

CALCULI: Kidney deposits formed as a result of abnormal states of the parathyroid gland, uric acid metabolism, and the like. Calculi, or "stones," may also be formed in the gall bladder, pancreas, and salivary glands.

CALYX: The green, clasping base of a flower (pl. calices).

CANDIDA ALBICANS: Formerly called *Monilia albicans*, this is a common, yeast-like fungus found in the mouth, vagina, and rectum, as well as on the outside skin; it is a common cause of thrush in infants and vaginal yeast infections. In recent years much attention has been given to the increased numbers of people who have developed candidiasis in the upper and lower intestinal tract. Formerly only found in extremely debilitated individuals, this condition can develop as a result of extended antibiotic therapy and anti-inflammatory treatment. Most anti-inflammatory drugs are really immunosuppressants, and this rather benign and common skin and mucosal fungus can move deeply into the body when resistance is drug-depressed, or the normal (and stable) competition between fungus and bacteria is altered in the treatment of bacterial infections. This condition is implicated in many allergies, mild auto-immunity conditions, and problems in sugar metabolism. Also considered by some of medicine to be just another type of trendy, quacky, "alternative" bit of silliness . . . like using herbs or seeing a chiropractor.

CAPILLARY: The smallest blood or lymph vessels, formed of single layers of inter-connected endothelial cells. They allow the transport across their walls and between their crevices of diffusible nutrients and waste products. Blood capillaries contract and expand, depending upon how much blood is needed to a given tissue and how much blood flows into them from the small arteries that feed them.

CARDIAC GLYCOSIDE: Sugar-containing plant substances that, in proper doses, act as stimulants to heart function. (Examples: digitalin, stropanthin.)

CARDIAC TONIC: A substance that strengthens or regulates heart metabolism without overt stimulation or depression, either by increasing coronary blood supply, normalizing coronary enervation, softening rigid arteries and decreasing backpressure on the heart's valves, or by decreasing unnecessary adrenergic stimulation. (Examples: Hawthorne berries, magnesium, Night Blooming Cereus.)

CATARRH: An inflammatory condition of any mucous membrane, usually accompanied by hypersecretion. The term is a little out-dated, but we still get nasal catarrh with a head cold.

CATHARTIC: An energetic laxative, often causing accompanying cramps, and usually needing an antispasmodic combined with it. Cathartics should not be used when there are inflammations or infections in the intestinal tract . . . too hot for "hot" conditions.

CENTRAL NERVOUS SYSTEM: A collective term for the brain, spinal cord, their nerves, and the sensory end-organs; more broadly, this may also include the neuro-transmitting hormones instigated by the CNS that control the chemical nervous system, the endocrine glands.

CERVICITIS: An inflammation, often infectious, of the cervix.

CERVIX: The neck and opening to the uterus.

CHOLECYSTITIS: An inflammation, usually induced by gallstones, of the gall bladder.

CHOLINERGIC: Literally, those nerve endings that secrete acetylcholine. The term refers more broadly to functions controlled by the parasympathetic nerves, such as their stimulation of digestive and intestinal secretions and muscles, constricting of the pupils, slowing the rate and strength of heartbeat, and the diminution and constriction of lung function . . . "yinny." Go sit under a tree, eat a meal, make love . . . cholinergic functions. To be fair, so are vomiting, diarrhea, and an asthma attack from inhaling pollen.

CHRONIC: A disease or imbalance of long, slow duration, showing little change. The opposite of acute.

CLUSTER HEADACHE: Migraine-like headaches that may occur several times a day for a week or two, usually with severe pain behind one eye. The causes are ill-understood but the results are often devastating; episodes are triggered by stress, trauma, overwork, allergies, or a blood-sugar shift.

COLIC: Spasms of the smooth muscles in any tube or duct. (Example: intestinal colic.)

COLITIS: An inflammation of the colon or large intestine, with or without infection.

COLOSTRUM: The fluid produced by the breasts immediately following birth and for several days afterwards before milk comes in. It contains antibodies and white blood cells and nutrients from the mother.

CONJUNCTIVITIS: Inflammation of the mucous membranes that line the eyelid and the eyeball.

CONSTITUTION: The basic physical and emotional profile of a person, including strengths and weaknesses that derive both from genetic patterns and patterns of lifestyle.

COROLLA: The petals or floral tube of a flower.

COUNTER-IRRITATION: The process of applying an irritating, heating, or vasodilating substance externally to the skin in order to speed healing locally by increasing circulation of fluids radiating the heat inward to inflamed tissues deep below the skin, or inducing reflex stimulation to seemingly unrelated internal organs. The latter use, ill-understood but useful, was formerly employed extensively by physicians but is rarely used in recent decades due to a lack of fundamental theory. It bears suspicious similarities to acupuncture and reflexology. (Examples: mustard plasters, moxibustion, firing.)

CROHN'S DISEASE: Also called regional enteritis or regional ileitis, this is a non-specific inflammatory disease of the upper and lower small intestine, forming granulated lesions. Usually forms a chronic condition, with acute episodes of diarrhea, abdominal pain, loss of appetite and weight. It may affect the stomach or colon, but the most common sites are duodenum and lower ileum. Standard treatment is, initially, anti-inflammatory drugs; surgical resectioning (sometimes several) may be necessary. The disease is probably auto-immune and sufferers share the same tissue type (HLA–B27) as those that acquire ankylosing spondylitis. (*See* Auto-Immunity.)

CYSTITIS: An inflammation of the urinary bladder, usually arising from a dystal infection of the urethra or prostate.

CYTOMEGALOVIRUS: This subtle, worldwide micro-organism is a member of the herpesvirus group. It is large (for a virus), contains DNA, and has a complex protein capsid. It forms latent, lifelong infections, and, except for occasional serious infections in infants and poor malnourished youngsters, seldom produces a disease state. With increased use of immunosuppression therapies for conditions ranging from arthritis to cancer to organ transplants, the incidence of adults with major infections arising from cytomegalovirus (CMV) increases yearly, as do the numbers of nonpatient infections. With herpes, Epstein-Barr virus, and candida, CMV is one of a number of endogenous, normally benign bugs that is afflicting what has jokingly been referred to as *Homo medicus*, that is, over-medicated members of industrialized societies.

DECIDUOUS: A plant that drops its leaves in the fall.

DECOCTION: A tea that is boiled slowly for fifteen or twenty minutes. *See* Format Explanation.

146

DEMULCENT: A substance that soothes the mucous membranes on contact. (Examples: Mallow, Hollyhock.)

DENTATE: A leaf margin that is regularly toothed like a saw.

DIABETES MELLITUS: A disease characterized by high blood sugar levels and sugar in the urine. Several different disorders that can be broken down into Juvenile Onset Diabetes (usually partially hereditary, with inadequate synthesis of native insulin, and therefore often termed Insulin-Dependent by physicians) and Adult-Onset Diabetes (partially inherited but mostly the result of lifestyle and diet, where hyperglycemic episodes have continued for so many years that fuel-engorged cells start to refuse excess glucose; often termed Insulin-Resistant by physicians).

DIAPHORETIC: A substance that increases perspiration. (Examples: Encelia, Marsh Fleabane.)

DIARRHEA: A watery evacuation of the bowels, without blood.

DISTAL: Away from the center.

DIURETIC: A substance that increases the flow of urine, either by increasing permeability of the kidney's nephrons or by increasing blood flow to the kidneys.

DIVERTICULOSIS: Usually congenital pouches found in many organs, particularly the colon, that are benign but prone to inflammation.

DOCTRINE OF SIGNATURES: An archaic theory holding that a plant will signify its use by a "signature." With this doctrine a plant with a kidney-shaped leaf is useful in treating kidney disorders. This form of sympathetic magic is still common in many societies.

DYSCRASIAS: Presently a term referring to improper synthesis by the liver of the various blood proteins, particularly in reference to clotting factors. Formerly used to describe an improper balance between blood and lymph in an organ or the whole body. More archaically, it referred to an imbalance between the four humours (blood, phlegm, black bile, and yellow bile).

DYSENTERY: Severe diarrhea, usually containing blood.

DYSPEPSIA: Poor digestion, usually with heartburn or regurgitation of stomach acids.

DYSPLASIA: Any abnormal tissue development.

DYSURIA: Painful urination, usually a sign of bladder, urethra, or prostate inflammation.

ECLECTICS: A former school of medicine, nearly extinct by the First World War. Eclectics were M.D.s trained under a curriculum aimed at low-tech, non-hospital rural health care. Because the Eclectics were trained for the use of less-invasive medicine (the country doc with a single black bag), they diverged radically from regular M.D.s in the first several decades of this century. They relied on what are now termed wholistic practices, and therefore maintained pharmacies that prepared botanical medicines for their use, long after regular doctors had pretty much dropped their use. Most schools with an Eclectic or "American" curriculum changed over to the high-tech hospital orientation outlined in the Flexner Report.

EDEMA: Local or general retention of excessive tissue fluids.

EMETIC: A substance that promotes vomiting.

EMMENAGOGUE: A substance that stimulates menstrual flow.

EMOLLIENT: A skin dressing or soothing ointment.

EMPHYSEMA: A chronic pulmonary disorder resulting from the loss of elasticity of the air spaces in the lungs, the coalescence of several spaces into one large one, and the decrease of surface area for gas interchange.

ENDOCRINE: Referring to the secretions that flow directly into the bloodstream. The various endocrine glands form a sort of chemical nervous system that is mutually interdependent.

ENDOMETRIUM: The inner mucous membrane of the uterus.

ENTIRE: A leaf with a straight, untoothed margin.

ENZYME: A protein catalyst, produced by living tissue, that speeds up the breakdown and simplification of other substances.

EPSTEIN-BARR VIRUS: A large, ubiquitous, and normally benign herpes-like virus, with both DNA and capsid, that has begun causing increasing amounts of disease among individuals having immunosuppressive therapy. It is sometimes implicated in mononucleosis and at least two types of lymphomas, as well as many subtle, poorly defined iatrogenic disorders. Recently re-named Chronic Fatigue Syndrome, or CFS. (*See also* Cytomegalovirus.)

ESTRONE: A physiologically active estrogen by-product.

ESTRUS CYCLE: The whole (usually) twenty-eight-day reproductive cycle in women.

ETHANOL: Ethyl alcohol, grain alcohol; e.g., the alcohol in booze.

EXOCRINE: A secretion that empties to the outside of the body, either to the external skin or the mucous membranes. The opposite of endocrine.

EXPECTORANT: A substance that stimulates the outflow of mucus from the lungs and bronchials.

FEBRILE: Feverish.

FERAL: Wild, native.

FIBROCYSTIC BREAST DISEASE: Normally, after ovulation, estrogen stimulates the growth of breast tissue and progesterone stimulates the alveolar secretions, both shrinking and draining during menses. In the most common form of FBD, the progesterone-induced secretions do not drain quickly enough back into the lymph, and fluid cysts gradually develop, becoming inflamed and tender a few days before each menses and eventually filling with varying amounts of fibrous tissue. For many women, FBD is aggravated by caffeine, theobromine, and theophylline. These substances occur, in various combinations, in coffee, green or black tea, cola nuts or cola drinks, maté, guarana, or chocolate.

FIXED OILS: Oils that leave a residue and do not evaporate. (Example: olive oil.)

FLATUS: Gas in the intestinal tract, either rising upwards or being expelled downwards.

FOMENTATION: A hot, wet poultice used on painful, inflamed areas. The usual form is a towel dipped in tea and applied hot or warm to the swollen area, being changed when it cools.

FUNGICIDE: A substance that kills or inhibits fungus infections. (Examples: Desert Willow, Cypress.)

G.I.: Referring to the gastro-intestinal tract.

GALENIC: Referring to the ancient physician Galen. Loosely applied to unrefined plant medications.

GASTRITIS: Inflammation of the stomach lining.

GASTROENTERITIS: Inflammation of the stomach and intestines; "stomach flu."

GIARDIASIS: Infection from Giardia lamblia, a flagellate protozoa that can cause diarrhea, dizziness, and weakness in some people, while no symptoms at all in others. Found in pristine snowfield runoff, all the way down to the humid tropics.

GLABROUS: A plant having no hairs.

GLAUCOMA: Disease of the eye, with increased intraocular pressure and eventual degeneration of the optic nerve and blindness. Usually chronic and slow in onset, aggravated by adrenergic stress habits and stimulant drugs.

GLAUCOUS: Hairless, but possessing a bloom, like Concord grapes.

GLYCOSIDE: A plant substance that, upon digestion, is absorbed into the bloodstream as a sugar plus one or more substances (aglycon) attached to it. (Examples: digitalin from Foxglove, arbutin from Manzanita.)

GLYCOSURIA: The presence of sugar (glucose) in the urine.

GOUT: A joint inflammation, usually of the big toe or knee, resulting from deposits of uric acid.

HEMATURIA: The presence of blood in the urine.

HEMOPHILUS: A genus of bacterium, most notable for needing the presence of certain blood proteins for their growth. Includes *H. influenza* (usually upper respiratory infections and pneumonia, as well as a type of meningitis and septicemia) and *H. vaginalis* (a common cause of sexually transmitted non-specific urethritis and non-specific vaginitis). Many cause secondary infections, as they are so common on healthy tissues.

HEMORRHOIDS: A mass of dilated veins in the anorectal area, either external and visible, or internal (piles). They may occur under two very different circumstances: athletic and physically active persons may get hemorrhoids from excessive anal sphincter tone ("jock hemmies"); or they may occur from friction, constipation, and generally poor tone of the pelvic veins and tissues. The two types usually respond to different therapies.

HEMOSTATIC: A substance that stops bleeding, either internally or externally. (Example: styptic pencil, Ratany, Shepherd's Purse.)

HEPATIC DUCT: The bile duct from the liver that bypasses the gall bladder and, with the gall bladder, empties into the common duct, thence into the duodenum. Hepatic bile drains the liver of waste products as well and is to the liver what urine is to the kidneys.

HEPATITIS: An inflammation of the liver. It may have many causes besides infectious hepatitis.

HEPATOCYTES: One of the three (and most common) types of functional cells of the liver, responsible for amino-acid synthesis and degradation, fat and lipid metabolization, and many other enzyme-mediated blood chemistry functions.

HEPATOSPLENOMEGALY: Enlargement of both the liver and spleen, most frequently the result of an insidious viral infection, such as viral hepatitis that has come to involve the spleen, or mononucleosis, which involves the liver and spleen.

HERPES SIMPLEX VIRUS: A small group of capsid-forming DNA viruses, divided into Type I (forming vesicles or blisters on the lips, mouth, eyes, and other areas more or less all above the waist) and Type II (usually sexually transmitted, with symptoms below the waist). Both types form acute initial outbreaks, go dormant, reactivate, go dormant, etc. For many, frequent outbreaks are clear symptoms of stress and immunosuppression. Both types are very dangerous to newborns and infants.

HIATUS HERNIA: An upwards protrusion of the stomach through the diaphragm wall, especially common in women in their fourth and fifth decades.

HOMEOPATHY: A substance approach to medicine. It works from the premise that something taken in large doses by a healthy individual produces symptoms in a certain order, chronology, and magnitude; therefore when highly diluted (attenuated), these same substances can be used in curing or at least alleviating those same symptom patterns in a sick individual. Although homeopathic medicines are by no means harmless, they are completely devoid of drug toxicity, and this practice is widespread throughout the world. In the United States perhaps only one hundred M.D.s are predominantly homeopathic. Only Arizona has a separate homeopathic licensing board. Traditionally, homeopathy acknowledges no diseases, only symptoms. The U.S. homeopathic pharmacopoeia is still official.

HYDROCELES: An accumulation of serous (lymph) fluids in a connective tissue sac, anywhere in the body, although the term is most frequently used to denote a fluid cyst of the testicle. Broadly, the internal equivalent of a blister.

HYGROSCOPIC: Having an affinity for, and readily absorbing, moisture.

HYPERGLYCEMIA: Elevations of blood glucose, either from the various types of diabetes or from adrenergic or stimulant drug causes.

HYPERURICEMIA: An abnormal amount of uric acid in the blood, usually above 7 mg. per 100 ml. of blood. This may lead to gout and certain types of kidney stones and may be the result of excess formation of uric acid or a hereditary tendency to poorly excrete it. Uric acid is the final waste product of the breakdown of nucleoproteins and, unlike urea, cannot be recycled.

HYPOGLYCEMIA: Abnormally low levels of blood glucose, usually transitory, and caused by fasting or impaired caloric intake. If the result of hyperinsulinemia (overproduction of pancreatic insulin) or too much insulin injected by a diabetic, it needs quick medical treatment. The hypoglycemia often referred to by nutritionists and alternative therapists is usually the labile blood sugar levels of the adrenergic, liver-deficient "yinny" person, whose glucose levels do a serum yo-yo. In the spirit of abject honesty, hypoglycemia, like candidiasis and parasites, can often be another pseudo-medical skyhook for some of us involved in alternative medicine and should not be confused with the strictly medical pathology of hypoglycemia.

IATROGENIC: Illness, disease, or imbalances, not present previously, created by medical or nonmedical treatment. In standard practice medicine, the therapy is blamed (not the therapist) and changed to something else. In alternative medicine, it may be called a "healing crisis" and deemed good for you. Beware: if the therapy makes you feel worse, it is almost always the wrong therapy.

IDIOPATHIC: A fancy term for "we don't know why you have it."

ILEO-CECAL VALVE: The sphincter that separates the ilium part of the (sterile) small intestine from the cecum, the ascending beginning of the (septic) colon. It protects the higher part of the intestines from the bacteria that are cultured in the colon and is heavily guarded by lymph nodes and Peyer's patches.

ILEUM: The lower two-thirds of the small intestine, ending in the ileo-cecal valve and emptying into the cecum of the colon. The last foot of the ileum is the only absorption site available for such important dietary substances as B-12, folic acid, some of the essential fatty acids, fat soluble vitamins, and recycled bile acids.

INFLUENZA: An acute respiratory infection caused by a large group of related viruses, highly contagious and most common in winter and spring ("flu season"). Every few years it puts on a camouflage suit of mutated proteins and makes the rounds of our species.

INFUSION: A method of brewing tea in which boiling water is poured over a plant. For specific types and processes, see Format Explanation.

IRRITABLE BOWEL SYNDROME (IBS): This is a common and benign condition of the colon, taking many different forms but characterized by alternating constipation and diarrhea. Usually there is some pain, particularly in the diarrhea phase. The main cause is stress, often with a history of several GI infections. The bowel equivalent of spasmodic asthma. Adrenergic stress slows the colon and causes constipation, followed by the cholinergic rebound overstimulation of the colon, with diarrhea. Also called spastic colon, colon syndrome, even chronic colitis.

JOCK ITCH: More elegantly called tinea cruris, it is a fungus infection of the groin, the crotch equivalent of athlete's foot.

LANCEOLATE: A leaf that is lance shaped.

LINOLEIC ACID: A polyunsaturated essential fatty acid, necessary in lipid and cholesterol metabolism as well as prostaglandin synthesis. Deficiency induces skin dryness and mucous membrane and conjunctival fragility and also contributes to excessive inflammatory responses.

LOCHIA: The cleansing drainage from the uterus following childbirth.

LIPID PEROXIDES: This term loosely refers to polyunsaturated fatty acids that have become "rancid" by contact with peroxide radicals, ozone, superoxide, hydroxyl radical, or singlet oxygen (O_1), decomposing into useless keto and hydroxy keto acids. The rancid smell of nuts, animal fats, and vegetable oils results partially from this breakdown and turns the important fatty acids into useless nutrients and metabolic damagers.

LYMPHADENITIS: Inflammation of lymph nodes, either as a defensive response to drainage into the node of infectious material distal to it (as a sore throat causing swelling of the nodes as they defend the throat) or from an infection of the nodes themselves (as in typhoid, mononucleosis, chlamydia).

LYMPHATIC: Pertaining to the lymph system or lymph tissue, the "back alley" of fluid circulation. Lymph is the alkaline, clear intercellular fluid that drains from the blood capillaries (where the arterial blood separates into thick, gooey veinous blood and lymph). It bathes the cells, drains up into the lymph capillaries, through the lymph nodes for cleaning and checking for bugs, up through the body and back to meet the veinous blood in the upper chest. Lymph nodes in the small intestine absorb most of the dietary fats. Lymph nodes and lymph tissue in the spleen, thymus, and tonsils also form lymphocytes and maintain the software memory of previously encountered antigens and their antibody defense response. The old axiom holds true: blood feeds the lymph, lymph feeds the tissues, lymph cleans the tissues and returns to the blood.

MACERATE: To soften and separate the parts of a substance by soaking for a length of time.

MARC: The insoluble remains of any extraction process: what you throw away when you are through.

MASTITIS: Inflammation of the breast, with or without infection.

MERIDIANS: Nonphysical energy pathways used in Oriental medicine to diagnose and stimulate internal or systemic imbalances. Meridian diagnosis plays a large role in traditional Oriental medicine, acupuncture, some forms of chiropractic, reflexology, and many approaches to bodywork and manipulation. They're real, but don't fit into present concepts of anatomy and physiology.

METABOLITE: Any by-product or waste product of metabolism.

MONONUCLEOSIS: Properly, infectious mononucleosis, a viral infection of the lymph tissue, characterized by enlarged spleen, lymph nodes, and often involving the liver.

MOXA: The prepared forms of several Artemisia species, either factory-rolled into a big paper-covered stick or used loose for rolling into little cones or balls. However used (and there are many forms in moxibustion), the moxa is lit and the ember provides heat that is applied therapeutically to parts of the body that have been diagnosed (usually by pulse or meridian diagnosis) as needing "dry heat."

MUCILAGE: Plant or animal starch characterized by thick, slimy viscosity, often used as emollient or demulcent agents. (Examples: Hollyhock, Mallow.)

MUCOPOLYSACCHARIDES: Starches (polysaccharides) that form chemical bonds with water. In the body, mucopolysaccharides lubricate joints, form the ground substance that holds cartilage together, and the starch hydrogel that keeps cells compacted into coherent masses and allows ready passage of lymph. Important therapeutic effects from plant mucopolysaccharides occur in Echinacea, Hollyhock, Prickly Pear, and other of the herbs in this book.

MUCOSA: The mucous membranes of the body, forming a continuous layer that protects it from the outside, where the outside has to go into or through the body, as in both the external-facing cavities (respiratory system and genitourinary system) and the intestinal tract. The inside skin of the body-doughnut.

NARCOTIC: A substance that depresses central nervous system function, bringing sleep and lessening pain. By definition, narcotics can be toxic in excess.

NATUROPATHY: A medical system that uses many and varied methods, including botanical medicines, to bring about the body's health by stimulating its innate defense mechanisms to establish balance and homeostasis. The basis of naturopathy is facilitating a "healing crisis" or return to health through nutrition, acupuncture, manipulation, herbs, hydrotherapy . . . whatever it takes. The practice almost disappeared in the 1950s, with unscrupulous individuals practicing or selling mail-order degrees, but has made a respectable comeback since that time with eight states (at this writing) licensing Naturopathic Physicians (N.D.s) and two four-year medical schools training them.

NEPHRITIS: An inflammation of the kidneys.

NERVINE: A substance that quells nervousness and irritability, either through depression or (occasionally) stimulation of the central nervous system. (Examples: Passion Flower, Wild Oats.)

NEURALGIA: Pain, sometimes severe, manifest along the length of a nerve and arising in the nerve itself, not in the tissue from which sensation is felt.

NEUROENDOCRINE: A term describing functions, substances, or responses that are integrated or shared by both the nervous systems and the endocrine glands. Epinephrine (adrenalin) is a neuroendocrine hormone, and pupil dilation is a neuroendocrine function.

OPPOSITE: Having the leaves arranged in pairs along opposite sides of the stem.

OVAL-LANCEOLATE: A leaf that is longer than wide but somewhat rounded in the center.

OXYTOCIC: A substance that stimulates uterine contractions and other related functions, either by mimicking the pituitary hormone oxytocin or by acting in synergy with it. (Examples: Pitocin, Cotton Root Bark, Shepherd's Purse.)

PALLIATIVE: Relieving symptoms without affecting the cause.

PALMATE: A hand-shaped leaf.

PARASYMPATHETIC: A division of the autonomic (involuntary) nervous system that controls normal digestive, reproductive, cardiopulmonary, and vascular functions, as well as stimulating secretions other than sweat. Sometimes called the "yin" influence of the nervous system, this subsystem works more or less as a direct antagonist with the other half of the autonomic system, the sympathetic. Organs balance between the parasympathetic and sympathetic systems. The term cholinergic refers to certain functions stimulated by the parasympathetic nerves.

PARASYMPATHOMIMETIC: A substance that mimics parasympathetic or cholinergic functions.

PAROTID GLANDS: These are the salivary glands that empty into the mouth by the upper second molars and give that sharp pleasant-pain sensation when you are about to eat brie cheese or drink dry wine after not eating for seven hours.

PARTURITION: Childbirth.

PERISTALSIS: The rhythmical, wavelike contractions of smooth muscles, especially those of the intestinal tract. They are controlled by the autonomic nerves (usually parasympathetic) and local nerve ganglia.

PHARYNGITIS: Inflammation of the pharynx; a sore throat.

PHENOLS: A group of aromatic benzene derivatives having bacteriostatic effects in small amounts, irritating and corrosive effects in large amounts.

PHYSIOLOGICAL DOSE: An amount of a therapeutic substance sufficient to cause a physical effect, as opposed to the minute quantities in attenuation of the substance when used homeopathically.

PHYTOSTEROL: A sterol, or fatty alcohol, found in many forms in plants.

PINNATE: A compound leaf having the leaflets arranged on either side of the leaf stem. (Example: Sweet Pea.)

PLEURISY: Inflammation of the membrane that enfolds the lungs.

PMS (Premenstrual Syndrome): A symptom picture manifested by some women frequently and most women occasionally, with water retention, breast engorgement, headaches, constipation, emotional irritability and/or depression, tendency towards binging. Any or all of these symptoms appear most frequently during the week before menses and usually end by the finish of menstruation. Sometimes it is attributed to imbalances between estrogen, progesterone, and prolactin arising after ovulation, but the factors are generally varied and poorly understood. PMS was formerly scoffed at by much of medicine; a sometimes "trendy" over-diagnosis these days, the reality probably lies between the two extremes.

PNEUMONITIS: Inflammation of the lungs; pneumonia.

PORTAL CONGESTION: The inability of the portal vein to drain properly into the liver, resulting in engorgement and impaired circulation in secondary, collateral veins, especially in the pelvis. Portal hypertension (above 30 cm. saline) is a more exaggerated condition, usually from cirrhosis, but simple congestion is a subclinical condition resulting from moderate engorgement of the liver sinusoids from poor protein metabolism, nutritional or metabolic lipid metabolism, alcohol or solvent abuse, and diet.

POULTICE: A hot, moist dressing placed over an injury, helping soothe pain, reduce congestive inflammation, and repair tissue damage, softening the resistance of the tissue to the fluids and blood below the surface, speeding up the reabsorption of fluids away from the injury, and sometimes even absorption of exudative fluids out of the injury and into the poultice by osmosis. For specific methods, see Format Explanation.

PROSTATITIS: In males, an inflammation of the prostate gland, a doughnut-shaped gland that surrounds the juncture of the urethra and the bladder and secretes semen fluids.

PROTEOLYTIC ENZYME: An enzyme that breaks down proteins.

PSORIASIS: A patchy dermatitis, usually of chronic nature and often inherited.

RACEME: An elongated, terminal flowering stem.

REFLEXED: Bent backwards, away from the direction of growth.

RENAL: Pertaining to the kidneys.

RESIN: A gummy, oily secretion or residue in or on a plant. (Example: Mesquite Gum.)

RHIZOME: A creeping horizontal root that gives rise to several stems.

RINGWORM: A fungal infection of the skin, forming red-ringed patches that itch and are mildly contagious. Synonym: tinea corporis, or, if on the scalp, tinea capitatis.

RUBIFACIENT: A substance that, when applied topically, dilates the blood vessels under the skin and brings heat to the area. If blistering occurs, it is termed a vesicant.

SAPONINS: Soapy glycosides found in the roots of some plants and usually having an irritating effect on the intestinal mucosa. (Example: Yucca saponins.)

SCIATICA: Neuralgia of the sciatic nerve, the largest in the body, that exits from the small of the back, descends down the back of each leg, splitting into two other major nerves (the peroneal and tibial). Sciatica is felt as severe pain from the buttocks, down the back of the thighs and usually radiating to the inside of the leg, sometimes with parasthesia (local numbing). It can arise out of lower back subluxation (responding to adjustment) or from pelvic congestion and edema (responding to laxatives, exercise, or relaxation of the pelvic and portal veins and better pelvic lymph drainage).

SEBACEOUS GLANDS: The oil-secreting glands of the skin, generally surrounding hair follicles.

SEBORRHEA: Disorder of the sebaceous glands, with changes in the amount and quality of the oils secreted. Although it can occur on many parts of the body, dandruff is probably the most common type.

SEDATIVE: Sleep inducing.

SEPAL: A leaf or segment of the calyx.

SEPTICEMIA: The presence of disease-causing bacteria in the bloodstream; serious and potentially fatal.

155

SEROUS MEMBRANES: Membranes inside the body, usually covering organs, that secrete serum as a lubricant. (Example: the pleura of the lungs.)

SHIGELLOSIS: An acute, self-limiting intestinal infection, with diarrhea, fever, and abdominal pain, caused by one of the *Shigella* genus of gram-negative bacteria. The infection is contracted through food prepared by infected individuals or by contact with them directly. Raw sewage may also be a source.

SITZ BATH: A cleansing or therapeutic bath in a small tub, holding enough water to sit in with the hips covered by the water and the legs and torso outside or above the water.

SKELETAL MUSCLES: Also called striped muscles or voluntary muscles, they are subject to conscious, central nervous system control.

SMOOTH MUSCLES: Also called involuntary muscles, they are stimulated (or suppressed) by the autonomic nervous system or peripheral ganglia and are not directly controlled by the central nervous system.

SPLEEN: The large organ lying to the left of, below, and behind the stomach; this organ is partially responsible for white blood cell formation (red in childhood), the filtering of the blood, removal and initial recycling of old and dead red blood cells, some of the resistance to infection, and the storage and concentration of red blood cells. Adrenalin and adrenergic enervation constricts the muscles that surround the spleen, ejecting the blood being processed back into circulation as a possible defense against injury or perhaps to increase the number of circulating blood cells for increased physical exertion.

STANDARD PRACTICE MEDICINE: The general body, at a given time, of acceptable diagnosis and therapeutics in the medical profession. Formerly largely derived from the procedures outlined in the Merck Manual and the more established specialists' texts, in recent years standard practice has come to mean those procedures covered within malpractice insurance guidelines. The consequence is that in rural, low population, and low income areas, standard practice entails less technology and less medical intervention . . . and lower malpractice rates. In higher income areas with higher malpractice premiums (insurance "redlining"), physicians, perhaps looking over their shoulders, utilize more high-tech diagnostics, more invasive or extensive intervention, and generally more expensive standard practice medicine. Understandable, perhaps, but insidious, nonetheless.

STEROIDS: Specifically, the hormones of the adrenal cortex and gonads or the various drugs, natural or synthetic, that mimic or supplant them; broadly, any number of lipids such as cholesterol, bile acids, vitamin D, and phytosterols.

STIPULE: A little appendage formed at the juncture of a leaf and the main plant stem.

SUBALPINE: Directly below timberline; high mountain forest.

SUPPURATING: Infected and discharging pus or, more broadly, sloughing off of tissue.

SUPRAINFECTION: Sometimes called a superinfection, this is a secondary infection by an organism that is resistant to antibiotics being given for a primary infection. Further, anti-inflammatory drugs and immunosuppressant therapy can, practically speaking, cause the same condition by inhibiting the patient's immunologic responses and allowing an organism to establish an infection that might otherwise not occur.

SYMPATHETIC: A division of the autonomic or involuntary nervous system that works in general opposition to the parasympathetic division. Much of the sympathetic functions are local and specific and secrete acetylcholine, like other nerves, to stimulate or suppress or whatever. A certain number of these nerves, however, unlike any others in the body, secrete epinephrine (adrenalin) and are called adrenergic. Since the adrenal medulla also secretes the same substances into the bloodstream as a hormone, all the muscles or glands that are affected by the adrenergic sympathetic nerves also react *in toto* to the epinephrine secreted by the adrenal medulla. This forms the basis for a potentially lifesaving emergency fight or flight response and is meant for short, drastic activities. A chronic excess of this adrenergic response, however, is a major cause of stress and sometimes can be a major cause of some chronic diseases. Since one of the first subjective symptoms of subclinical malnutrition and metabolic imbalances is irritability, hypersympathetic functions act as an intermediate between poor diet and disease.

SYMPATHOMIMETIC: A substance that mimics adrenalin (epinephrine). (Examples: ephedrine, amphetamines, caffeine, cocaine.)

SYSTEMIC: Reacting on the body as a whole rather than on a single tissue or organ, and, by inference, carried in the bloodstream.

SYSTOLIC: The measurement of arterial blood pressure at the point of heart contraction (greatest pressure), as opposed to diastolic, which is the measurement at rest (least pressure).

TACHYCARDIA: Abnormally fast heart beat.

TANNINS: Protective substances found in the outer and inner tissues of plants, breaking down through time to phlebotannins and, finally, humin. All are relatively resistant to digestion or fermentation, and either decrease the ability of animals to easily consume the living plant or, as in deciduous trees, cause the shed parts of the plant to decay so slowly that there is little likelihood of infection to the still-living parts of the plant resulting from rotting dead material around its base. All tannins act as astringents, shrinking tissues and contracting structural proteins in the skin and mucosa, but differ in their chemistry and accompanying substances. Tannin-containing plants can vary a great deal in their physiological effects and should not be lumped together, as they often are.

TENDONITIS: Inflammation of the sinew that attaches a muscle to a bone.

TENESMUS: Painful spasmodic expelling contractions of intestinal muscles, usually of the large intestine (colon), rectum, and anus.

THOMSONIAN: That school of medical philosophy and therapy founded by Samuel Thomson (b. 1769). Thomson's great axiom was "Heat is life, and cold is death." He and the later Thomsonians made great use of vomiting, sweating, and purging to achieve these ends . . . crude by present standards, but saner than the standard practice medicine of the times. The Thomsonians split vehemently from the early Eclectics before the Civil War; the latter, larger group preferred to train physicians as M.D.s, the first group disavowed any overt medical training . . . "physicking." Many of the practices of Jethro Kloss (*Back to Eden*) are Thomsonian, and John Christopher could be termed a Neo-Thomsonian. *Rule of Thumb:* If you see Lobelia and Capsicum together in a formula, it's probably something Sam Thomson did first.

TINCTURE: A water and alcohol concentration of a plant, used either for convenience (for an external liniment or for an easily made tea) or because some of the active constituents are not very soluble in water. For procedures see Format Explanation.

TINEA: Dermatomycosis; any number of topical fungus infections, such as ringworm (tinea capitis if on the head), athlete's foot, jock itch, etc.

TONIC: A substance taken to strengthen and prevent disease, especially chronic. Formerly widely available both as over-the-counter and prescription formulas, the increased sophistication of medicine in the last several decades has led to predominantly disease-at-a-time treatment, with almost no current use of a tonic or preventative approach . . . hence one of this author's functions.

TONSILLITIS: Inflammation of the tonsils, especially those in the throat (the lymphatic pulp organs) that respond and defend against mouth and throat infections before they can move deeper into the body.

TOXEMIA: Autotoxicity, either through septicemia and the systemic distribution of bacterial exotoxins or through the breakdown of some excretory functions and the resultant recycling of waste products in the blood.

TOXINS: Poisonous substances of an external or internal nature. The disease reactions to many microbes are often due to their toxic waste products and not the presence, per se, of the parasite. Fever, lethargy, and inflammations are some defensive responses that the body can manifest in the presence of excess toxin levels.

TRIMESTER: The three three-month divisions of a nine-month pregnancy.

TUBER: The thick, starchy underground stems or roots of many plants, often with buds present.

UMBEL: A flower or seed formation in which the stems radiate to each flower like the supports in an upended umbrella. (Example: Celery.)

URETHRA: The canal between the bladder and the outside of the body.

URETHRITIS: Any inflammation of the urethra, whether from external irritation, overly acidic or scalding urine, or an active infection of the canal.

URIC ACID: The final end product of certain proteins in the body or from the diet, especially the nucleoproteins found in the nucleus of cells. Unlike the much smaller protein waste product, urea, which is mostly recycled to form many amino acids, uric acid is an unrecyclable metabolite, a bent nail that won't restraighten and must be excreted . . . nucleoprotein to purine to uric acid to the outside in the urine or sweat.

VAGINITIS: An inflammation of the vagina.

VASOCONSTRICTOR: A substance that causes a narrowing or constriction of (especially) the smaller peripheral arteries.

VASODILATOR: A substance that causes a relaxing and dilation of blood vessels, and, usually, less back-up pressure on the large arteries and the heart.

VERMICIDE: A substance that kills or expels intestinal worms.

VOLATILE OILS: Those oils that are distillable and will evaporate in the air and leave no residue.

VESICULAR STOMATITIS: Also called aphthous stomatitis, it is an acute inflammation with many little blisters on the mouth, tongue, or lips. The cause can range from an overt infection to food allergies to irritants like tobacco and chile to chronic heartburn.

VARICOSITIES: The presence of distended, twisted, or swollen veins or venules (small veins that drain the blood capillaries).

XEROPHYTE: A plant that is drought resistant or is found in the deserts.

SELECTED REFERENCES AND FURTHER READING

Botany

ARNBERGER, L. P., AND J. R. JANISH. *Flowers of the Southwest Mountains.* Southwest Parks and Monuments Assn., Tucson, Ariz., 1982.

BENSON, LYMAN, AND ROBERT A. DARROW. *Trees and Shrubs of the Southwestern Deserts.* 3rd ed. University of Arizona Press, Tucson, 1981.

CORRELL, D. S., AND M. C. JOHNSON. *Manual of the Vascular Plants of Texas.* Texas Research Foundation, Renner, 1970.

DODGE, NATT. *100 Roadside Wildflowers of the Southwest Uplands.* Southwest Parks and Monuments Assn., Tucson, Ariz., 1967.

———. *100 Desert Wildflowers.* Southwest Parks and Monuments Assn., Tucson, Ariz., 1963.

DODGE, NATT, AND J. R. JANISH. *Flowers of the Southwest Deserts.* Southwest Parks and Monuments Assn., Tucson, Ariz., 1985.

EARLE, W. HUBERT. *Cacti of the Southwest.* 2nd ed. W. Hubert Earle, Phoenix, Ariz., 1980.

ELMORE, F. H. AND J. R. JANISH. *Shrubs and Trees of the Southwest Uplands.* Southwest Parks and Monuments Assn., Tucson, Ariz., 1976.

IVEY, ROBERT D. *Flowering Plants of New Mexico.* Robert D. Ivey, Albuquerque, 1983.

JAEGER, EDMUND. *Desert Wild Flowers.* Stanford University Press, Stanford, Calif., 1956.

JEPSON, WILLIS. *Manual of the Flowering Plants of California.* University of California Press, Berkeley, 1960.

KEARNEY, THOMAS, AND ROBERT PEEBLES. *Arizona Flora.* University of California Press, Berkeley, 1964.

LAMB, SAMUEL H. *Woody Plants of New Mexico.* New Mexico Department of Game and Fish Bulletin, Santa Fe, 1971.

MARTIN, W. C., AND C. R. HUTCHINS. *A Flora of New Mexico.* 2 vols. J. Cramer, Lehre, Germany, 1980.

MASON, CHARLES, AND PATRICIA MASON. *A Handbook of Mexican Roadside Flora.* University of Arizona Press, Tucson, 1987.

MUNZ, PHILIP A. *A California Flora.* University of California Press, Berkeley, 1968.

———. *A Flora of Southern California.* University of California Press, Berkeley, 1974.

———. *California Desert Wildflowers.* University of California Press, Berkeley, 1963.

PARKER, KITTIE F. *An Illustrated Guide to Arizona Weeds.* University of Arizona Press, Tucson, 1972.

PATRAW, P. M., AND J. R. JANISH. *Flowers of the Southwest Mesas.* Southwest Parks and Monuments Assn., Tucson, Ariz., 1973.

RICKETT, HAROLD W. *Wild Flowers of the United States.* Volume 4. *The South-western States.* McGraw Hill Book Company, New York, 1973.
―――. *Wild Flowers of the United States.* Volume 6. *The Central Mountains and Plains.* McGraw Hill Book Company, New York, 1973.
SHREVE, FORREST, AND IRA WIGGINS. *Vegetation and Flora of the Sonoran Deserts.* 2 vols. Stanford University Press, Stanford, Calif., 1964.
SPELLENBERG, RICHARD. *The Audubon Society Field Guide to North American Wildflowers, Western Region.* Alfred A. Knopf, N.Y., 1979.
VINES, ROBERT A. *Trees, Shrubs and Woody Vines of the Southwest.* University of Texas Press, Austin, 1960.
WIGGINS, IRA. *Flora of Baja California.* Stanford University Press, Stanford, Calif., 1980.
WILLS, M. M., AND H. S. IRWIN. *Roadside Flowers of Texas.* University of Texas Press, Austin, 1961.

Herbal Use and Ethnobotany

BALLS, EDWARD K. *Early Uses of California Plants.* University of California Press, Berkeley, 1962.
BURLAGE, HENRY M. *Index of Plants of Texas with Reputed Medicinal and Poisonous Properties.* Burlage, Austin, 1968.
CHAVEZ, TIBO J. *New Mexican Folklore of the Rio Abajo.* Bishop Printing Co., Portales, 1972.
CULBRETH, DAVID. *A Manual of Materia Medica and Pharmacology.* Eclectic Medical Publications, Portland, Oreg., 1983 reprint.
CURTIN, L. S. M. *By the Prophet of the Earth.* San Vicente Foundation, Santa Fe, N.M., 1949.
―――. *Healing Herbs of the Upper Rio Grande.* Southwest Museum, Los Angeles, Calif., 1965.
ELLINGWOOD, FINLEY. *American Materia Medica.* Eclectic Medical Publications, Portland, Oreg., 1983 reprint.
FELGER, R. S., AND M. B. MOSER. *People of the Desert and Sea.* University of Arizona Press, Tucson, 1985.
FELTER, H. WICKS. *Eclectic Materia Medica.* Eclectic Medical Publications, Portland, Oreg., 1983 reprint.
―――, AND JOHN U. LLOYD. *King's American Dispensatory.* 2 vols. Eclectic Medical Publications, Portland, Oreg., 1983 reprint.
FORD, KAREN. *Las Yerbas de la Gente.* Museum of Anthropology, University of Michigan, Ann Arbor, 1975.
GREGORY, MICHAEL. *Bibliography Preliminary to a Western Herbal.* Bisbee, Ariz., 1978.
GRIEVE, MAUD. *A Modern Herbal.* 2 vols. Dover Publications, N.Y., 1971 reprint.
HARPER-SHOVE, F. *Prescriber and Clinical Repertory of Medicinal Herbs.* Health Science Press, Devon, England, 1952.
HOCKING, GEORGE M. *A Dictionary of Terms in Pharmacognosy.* Thomas, Springfield, Ill., 1955.

KROCHMAL AND KROCHMAL. *A Field Guide to Medicinal Plants.* Times Books, N.Y., 1984.

KUTS-CHERAUX, A. W. *Naturae Medicina and Naturopathic Dispensatory.* American Naturopathic Physicians and Surgeons Assn., Des Moines, Iowa, 1953.

MARTINDALE'S *Extra Pharmacopea.* 25th ed. Pharmaceutical Press, London, England, 1967.

MARTINEZ, MAXIMO. *Las Plantas Medicinales de Mexico.* 4th ed. Botas, Mexico City, 1959.

MEYER, JOSEPH. *The Herbalist.* Indiana Botanic Gardens, Hammond, n.d.

MILLSPAUGH, CHARLES. *American Medicinal Plants.* Dover Publications, N.Y., 1974.

MOORE, MICHAEL. *Los Remedios de la Gente.* Herbs Etc., Santa Fe, N.M., 1977.

————. *Medicinal Plants of the Mountain West.* Museum of New Mexico Press, Santa Fe, 1979.

————. *Herbal Repertory in Clinical Practice.* School of Clinical Herbology, Santa Fe, N.M., 1983.

OLIVER, J. H. *Proven Remedies.* Thorsons, London, England, 1962.

POWELL, ERIC F. *The Modern Botanical Prescriber.* L. N. Folwer, London, England, 1965.

PRIEST AND PRIEST. *Herbal Medication, a Clinical and Dispensary Handbook.* L. N. Fowler, London, England, 1982.

SPINELLI, WILLIAM B. *The Primitive Therapeutic Use of Natural Products, a Bibliography.* Duquesne University, Pittsburgh, Pa., 1971.

STEINMETZ, E. F. *Materia Medica Vegetabilis.* E. F. Steinmetz, Amsterdam, Netherlands, 1954.

————. *Drug Guide for Importers, Exporters, Dealers, Etc.* E. F. Steinmetz, Amsterdam, Netherlands, 1959.

UPHOF, J. C. *Dictionary of Economic Plants.* 2nd ed. J. Cramer, Lehre, Germany, 1968.

VOGEL, VIRGIL. *American Indian Medicine.* University of Oklahoma Press, Norman, 1970.

WILSON, CLOYCE. *Useful Prescriptions.* Lloyd Brothers Pharmacy, Cincinnati, Ohio, 1935.

WREN, R. C. *Potter's New Cyclopedia of Botanical Drugs and Preparations.* Harper and Row, N.Y., 1970.

Homeopathy

BOERICKE, WILLIAM. *Materia Medica with Repertory.* Boericke and Tafel, Philadelphia, Pa., 1927.

CLARKE, JOHN H. *A Dictionary of Materia Medica.* 3 vols. Health Science Press, Sussex, England, 1962 reprint.

Homeopathic Pharmacopoeia of the United States. 7th ed. Boericke and Tafel, Philadelphia, Pa., 1964.

Medical and Pharmaceutical

GOODHART, R. S., AND M. E. SHILS. *Modern Nutrition in Health and Disease.* 5th ed. Lea and Febiger, Philadelphia, 1973.

Gray's Anatomy of the Human Body. 29th ed. Lea and Febiger, Philadelphia, 1973.

The Merck Index. 9th ed. Merck and Co., Rahway, N.J., 1976.

The Merck Manual. 13th ed. Merck and Co., Rahway, N.J., 1977.

MILLER, L. P., ed. *Phytochemistry.* Van Nostrand Rheinhold Co., N.Y., 1973.

ROBINSON, T. *The Organic Constituents of Higher Plants.* Burgess Publ., Minneapolis, 1967.

ROTHENBERG, R. E. *Medical Dictionary and Health Manual.* Signet Books, N.Y., 1975.

STEEN, EDWIN, AND ASHLEY MONTAGU. *Anatomy and Physiology.* 2 vols. Barnes and Nobel, N.Y., 1959.

Taber's Cyclopedic Medical Dictionary. 13th ed. F. A. Davis, Philadelphia, 1977.

TREASE, G. E., AND W. C. EVANS. *A Textbook of Pharmacognosy.* 10th ed. Tindall and Cassell, London, England, 1972.

TYLER, VARRO E., LYNN R. BRADY, AND JAMES E. ROBBERS. *Pharmacognosy.* 7th ed. Lea and Febiger, Philadelphia, 1976.

TYLER, VARRO E., AND A. E. SCHWARTING. *Experimental Pharmacology.* 3rd ed. Burgess Publ., Minneapolis, 1962.

WALLIS, T. E. *Textbook of Pharmacognosy.* 5th ed. J and A Churchill, London, England, 1967.

Wildflower Horticulture, Aesthetics

BUBEL, NANCY. *The Seed-Starter's Handbook.* Rodale Press, Emmaus, Pa., 1978.

———. *Landscaping with Native Arizona Plants.* University of Arizona Press, Tucson, 1973.

LENZ, LEE W. *Native Plants for California Gardens.* Rancho Santa Ana Botanical Gardens, Claremont, Ca., 1973.

NABHAN, GARY P. *The Desert Smells Like Rain.* University of Arizona Press, Tucson, 1982.

———. *Gathering the Desert.* University of Arizona Press, Tucson.

NOKES, JILL. *How to Grow Native Plants of Texas and the Southwest.* Texas Monthly Press, Austin, 1986.

PHILLIPS, JUDITH. *Southwestern Landscaping with Native Plants.* Museum of New Mexico Press, Santa Fe, 1987.

TAYLOR, K. S., AND S. F. HAMBLIN. *Handbook of Wildflower Cultivation.* Macmillan Publishing Co., N.Y., 1963.

WHEALY, KENT. *The Garden Seed Inventory.* Seed Saver Publishers, Decorah, Iowa, 1985.

THERAPEUTIC AND USE INDEX

This is an index only. Please refer to the main text for specific information before using any herb. Formulas are in italics.

ABORTIVE
Cotton Root (?)
Epazote (?)

ALCOHOLISM
General:
Escoba de la Vibora
Milk Thistle
Recuperation:
Indian Root
Milk Thistle
Alterative Syrup (#15)
Hangovers/Acute Gastritis:
Acacia
Desert Lavender
Sage
Tronadora
Turkey Mullein (external)
Chronic Gastritis/Pacreatitis:
Algerita
Prodigiosa

ALLERGIES
General Support:
Algerita
Chaparral
Indian Root
Prodigiosa
Stillingia (external)
Hay Fever:
Canadian Fleabane
Condalia
Eucalyptus
Prickly Pear
Red Root
Sage
Yerba Mansa
(*See also:* Auto-Immune
Diseases, Eyes, Liver,
Metabolic Disorders,
Respiratory, Sinuses.)

ALTERATIVE
Algerita
Chaparral
Echinacea
Indian Root
Milk Thistle
Ocotillo
Stillingia
Alterative Syrup (#15)

ANTI-INFLAMMATORY
General, systemic:
Soapberry
Yerba Mansa
Mucous Membranes:
Jojoba
Mallow
Topical:
Cañaigre Leaves
Pineapple Weed
Sangre de Drago
Emollient Poultice (#9)

ANTIFUNGAL
General, external:
Camphor Weed
Chaparral
Cypress
Desert Willow
Epazote
General, internal:
Desert Willow
Echinacea
Stillingia
Candidiasis:
Desert Willow
Echinacea
Eucalyptus (upper G.I.)
Stillingia
Trumpet Creeper
Alterative Syrup (#15)

ANTIMICROBIAL/ ANTISEPTIC
External:
  Acacia
  Algerita
  Camphor Weed
  Chaparral
  Condalia
  Cypress
  Desert Willow
  Epazote
  Eucalyptus
  Juniper
  Mesquite
  Piñon
  Sage
  Yerba Mansa
Internal:
  Algerita
  Chaparro Amargosa
  Desert Willow
  Echinacea
  Elephant Tree
  Mesquite
  Yerba Mansa

ANTIOXIDANT
  Chaparral
  Piñon

ANTISPASMODIC
  Marsh Fleabane
  Passion Flower
  Pineapple Weed
  Prickly Poppy
  Silk Tassel
  Turkey Mullein
  Vervain
  Wild Lettuce
  Yerba Mansa
(*see* Muscle/Joint Pain,
Sedative/Tranquilizer.)

ANTIVIRAL
  Echinacea
  Tronadora
  Immunostimulant (#8)

ARTHRITIS
General, external:
  California Mugwort
  Camphor Weed
  Chickweed
  Escoba de la Vibora
  Matarique
  Pineapple Weed
  Prodigiosa
  Yerba Mansa
  Yerba de A. Garcia
(*See also* Counter-irritant,
Linament, Poultice
Anti-inflammatory, internal:
  Encelia (acute)
  Escoba de la Vibora
  Soapberry (acute)
  Yerba Mansa
  Yucca
Diuretic (with faulty fat and
protein metabolism):
  Agave
  Chaparral
  Echinacea
  Shepherd's Purse
  Yerba Mansa
  Alterative Syrup (#15)
(*See* Auto-Immune, Gout,
Muscle/Joint Pain.)

ASTRINGENT
Topical:
  Acacia
  Buckwheat Bush
  Cañaigre
  Condalia
  Elephant Tree
  Mesquite
  Ratany
  Sage
  Sangre de Drago
  Sumach
Internal:
  Canadian Fleabane
  Cypress (roasted)
  Mesquite

Ratany
Sangre de Drago
(*See* Capillary Fragility,
Hemostatic, Skin.)

AUTOIMMUNE DISEASES

Supportive Therapy:
Chaparral
Echinacea
Elephant Tree
Milk Thistle
Yerba Mansa
Alterative Syrup (#15)

BACKACHE

General, in bath:
Camphor Weed
Escoba de la Vibora
Yerba de Alonso Garcia
Vague, nagging, not localized:
Cliff Rose
Silk Tassel
(*See* Muscle/Joint Pain.)

BATH/SAUNA HERBS

Escoba de la Vibora
Eucalyptus
Juniper
Sage (White)
Sagebrush
Turkey Mullein
Yerba Mansa
Yerba de Alonso Garcia

CAPILLARY FRAGILITY

General:
Prickly Pear Flowers
Red Root
Shepherd's Purse
Mucous Membranes:
Jojoba
Chronic Bleeding Formula (#3)
(*See* Astringent, Hemostatic.)

CARDIAC

Palpitations (from food, hormonal changes):
Prickly Poppy
Palpitations (from stress, anxiety):
Turkey Mullein
Palpitations (with tachycardia in smokers, caffeine users):
Night Blooming Cereus
Rapid, wiry pulse (with high blood pressure):
Syrian Rue
Rapid, wiry pulse (with low vitality):
Anemone
(*See* Cardiovascular, Hypertension, Stress.)

CARDIOVASCULAR

Arteriosclerosis:
Puncture Vine
Pelvic Congestion:
Ocotillo
Adrenergic Stress (peripheral vasoconstriction, dry G.I.):
Indian Root
Periwinkle
(*See* Cardiac, Hypertension, Metabolic, Stress.)

CHEWING GUM

Elephant Tree
Piñon

COLDS/FLU

California Mugwort
Cenizo
Echinacea
Encelia
Tronadora
Vervain
Yerba Mansa
Yerba de Alonso Garcia
Immunostimulant (#8)
(*See* Allergies, Cough,

Respiratory, Throat
Inflammation.)

COLIC, Children
Chimaja
Desert Lavender
Mallow
Passion Flower
Pineapple Weed
Vervain
Wild Lettuce
(*See* Pregnancy/PostPartum,
Stomach.)

CONVALESCENT AID
General:
Indian Root
Spikenard
Alterative Syrup (#15)
Intestinal:
Mesquite
Intestinal Disinfectant (#10)
Liver:
Milk Thistle
Lipid Normalizer (#2)
Respiratory:
Stillingia
Spikenard
Alterative Syrup (#15)

COUGH
Dry, raspy, hot:
Acacia
Hollyhock
Cliff Rose
Mallow
Dry, raspy (with flushed face,
rapid pulse):
Desert Willow
Immunostimulant (#8)
with Sleeplessness
Wild Lettuce
Recuperating from acute:
Encelia
Piñon
Stillingia
Alterative Syrup (#15)

Moist, chronic, tired:
Canadian Fleabane
Ocotillo
Piñon
Spikenard
with Muscle Pains:
Turkey Mullein
Smoker's Cough:
Spikenard
Stillingia
(*See* Colds/Flu, Respiratory,
Throat Inflammation.)

COUNTERIRRITANT
Camphor Weed
Condalia
Piñon
Prickly Poppy
Turkey Mullein
(*See* Arthritis, Linament Herbs.)

CROHN'S DISEASE
General Support:
Milk Thistle
Yerba Mansa
Chronic Bleeding Formula
(#3)
For Pain:
Silk Tassel
(*See* Autoimmune Disease,
Intestinal Tract.)

CYTOMEGALOVIRUS
(CMV)
Milk Thistle with Red Root
Alterative Syrup (#15)

DEFOLIANT
Chaparral

DEMULCENT/
EMOLLIENT
Acacia
Hollyhock
Mallow
Pineapple Weed
Emollient Poultice (#9)
(*See* Poultice Herbs.)

DIABETES
Adult Onset, insulin-
resistant:
  Matarique
  Prickly Pear
  Prodigiosa
  Tronadora
(*See* Hyperglycemia,
Metabolic Disease.)

DIAPHORETIC
General:
  California Mugwort
  Cenizo
  Cliff Rose
  Encelia
  Eucalyptus
  Indian Root
  Marsh Fleabane
  Sagebrush
  Silver Sage
Chills (with fever):
  Sage
Dry Fevers, early on:
  Turkey Mullein
  Vervain
Antidiaphoretic:
  Canadian Fleabane
  Sage
(*See* Fevers.)

DIURETIC
General:
  Buckwheat Bush
  Eucalyptus
  Marsh Fleabane
  Vervain
Sodium Loss, anabolic stress:
  Chickweed
  Puncture Vine
  Shepherd's Purse
  Yerba Mansa
Premenstrual Water
Retention:
  Chickweed

  Shepherd's Purse
Water Retention (with dry
skin):
  Silver Sage
Water Retention (from
weather changes):
  Agave
  Juniper
  Puncture Vine
Antidiuretic:
  Canadian Fleabane
(*See* Kidneys, Urinary Tract.)

DOUCHE
Astringent, soothing:
  Manzanita
  Sumach
  Mallow
Antimicrobial:
  Hollyhock
  Mesquite
  Trumpet Creeper
  Yerba Mansa
Cervical erosion:
  Echinacea
  Sangre de Drago
(*See* Reproductive System—
Female.)

DRUG ABUSE
Cocaine:
  Wild Oats
  Sumach
  Yerba Mansa
Hallucinogenics, anxiety
reaction:
  Anemone with Passion
  Flower
Hallucinogenics, coming
down:
  Wild Oats
(*See* Alcoholism, Psychiatric,
Nose.)

ENDOCRINE SYSTEM
Thyroid:
  Algerita

Oxytocin Synergist:
  Cotton Root
  Shepherd's Purse
Steroidal:
  Agave
  Hollyhock
  Yucca
(*See* Reproductive System—
Female, Male, Stress.)

ENEMA
  Desert Willow (candidiasis)
  Sumach
  Yerba Mansa

EPSTEIN-BARR
VIRUS (EBV)
  Milk Thistle with Red Root
  Alterative Syrup (#15)
(*See* Cytomegalovirus,
Immunologic.)

EXPECTORANT
General:
  Condalia
  Juniper
  Piñon
Disinfecting:
  Elephant Tree
  Eucalyptus
  Juniper
To Moisten (in early acute):
  California Mugwort
To Soften (in later, subacute):
  Cliff Rose
  Encelia
  Elephant Tree
  Eucalyptus
  Piñon
To Dry Out (from allergies or
chronic bronchorrhea):
  Canadian Fleabane
(*See* Cough, Respiratory.)

EYES
Conjunctivitis Wash:
  Acacia

  Buckwheat Bush
  Condalia
  Marsh Fleabane
  Mesquite
Cataracts:
  Prodigiosa (?)

FEVERS
Periodic or night:
  Soapberry
  Night Sweat Formula (#14)
Shaking, chills:
  Sage
  Turkey Mullein
with Upset Stomach, aches:
  Vervain
Suppress:
  Yerba Mansa
(*See* Colds/Flu,
Diaphoretics)

FISHING
  Soapberry
  Turkey Mullein

FLYPAPER
  Chaparro Amargosa
  Flypaper (#5)

FOOD
  Cañaigre
  Chimaja
  Condalia
  Manzanita
  Mesquite
  Passion Flower
  Prickly Pear
  Prickly Poppy
  Sagebrush
(*See* Spices, Teas.)

FREE RADICAL
THERAPY
  California Mugwort
  Chaparral
  Enchinacea
  Milk Thistle

(*See* Anti-oxidant,
Autoimmune Disease.)

GOUT
Shepherd's Purse
Yerba Mansa
(*See* Arthritis,
Hyperuricemia, Metabolic,
Stress.)

HAIR
Dandruff/Seborrhea:
  Agave
  Condalia
  Jojoba
  Red Root
  Soapberry
  Syrian Rue
Rinse/Conditioner:
  Jojoba
  Sage
(*See* Soap/Shampoo.)

HEADACHE
Sick, nauseous, frontal:
  Buckwheat Bush
  California Mugwort
Migraines:
  Periwinkle
with Bloodshot Eyes,
Nausea:
  Silver Sage (wash)
  Turkey Mullein (wash

HEMORRHOIDS
Acute:
  Añil del Muerto
  Desert Lavender
  Periwinkle
  Ratany
  Sangre de Drago
Subacute:
  Canadian Fleabane
  Jojoba
  Red Root
Chronic with congestion:
  Milk Thistle

  Ocotillo
Ointment:
  Echinacea
  Hemorrhoid Ointment (#1)

HEMOSTATIC
General Topical:
  Acacia
  Cañaigre
  Ratany
  Sangre de Drago
Internal:
  Canadian Fleabane
  Desert Lavender
  Periwinkle
  Ratany
  Red Root
  Shepherd's Purse
  Vervain
  Chronic Bleeding Formula
  (#3)

HYDROCELES/FLUID CYSTS
Red Root in combination (see
text).

HYPERGLYCEMIA
  Matarique
  Prickly Pear
  Prodigiosa
  Tronadora
(*See* Diabetes, Stress.)

HYPERLIPIDEMIA/ HYPERCHOLES-TEROLEMIA
  Chaparral
  Milk Thistle
  Prickly Pear
  Puncture Vine
  Lipid Normalizer (#2)

HYPERTENSION
Essential, in strong

middle-aged:
  Passion Flower
  Puncture Vine
  Hypertension Formula (#11)
Episodic, mild (without
medications):
  Periwinkle
with Tachycardia, physically
tired:
  Syrian Rue
(*See* Cardiac,
Cardiovascular.)

HYPERURICEMIA
  Chickweed
  Puncture Vine
  Shepherd's Purse
  Yerba Mansa
(*See* Gout.)

IMMUNOLOGIC
General Stimulant:
  Cypress
  Echinacea
  Elephant Tree
  Hollyhock
  Indian Root
  Ocotillo
  Spikenard
  Yerba Mansa
  Immunostimulant (#8)
Leukocyte Stimulant:
  Echinacea
  Elephant Tree
  Hollyhock
  Indian Root
  Spikenard
Opsonization & Complement
Stimulant:
  Echinacea
(*See* Autoimmune, Lymph.)

INCENSE
  Cypress
  Encelia
  Elephant Tree
  Juniper

  Piñon
  Sage (White)
  Sagebrush
  Spikenard

INSECT REPELLENT
  Cypress (clothes)
  Echinacea
  Sagebrush

INTESTINAL TRACT
Colitis:
  Canadian Fleabane
  Hollyhock
  Jojoba
  Mesquite
  Prickly Pear
  Ratany
  Shepherd's Purse
  Silk Tassel
  Diarrhea Formula (#13)
Cramps, Tenesmus:
  Marsh Fleabane
  Matarique
  Passion Flower
  Pineapple Weed
  Silk Tassel
  Turkey Mullein
  Vervain
  Wild Lettuce
(*See* Antispasmodic.)
Diarrhea/Dysentery:
  Acacia
  Canadian Fleabane
  Cañaigre
  Chaparro Amargosa
    (protozoan)
  Cypress
  Hollyhock
  Mesquite
  Passion Flower (pain)
  Prickly Poppy
  Ratany
  Sangre de Drago
  Silk Tassel (pain)

Turkey Mullein (bleeding)
Intestinal Disinfectant (#10)
Diarrhea Formula (#13)
Diverticulitis:
  Canadian Fleabane
  Prickly Pear
  Yerba Mansa
Inflammation, acute:
  Acacia
  Cypress
  Echinacea
  Hollyhock
  Mesquite
  Ratany
  Sangre de Drago
Inflammation, chronic:
  Echinacea
  Jojoba
  Ocotillo
  Yerba Mansa
  Chronic Bleeding Formula
    (#13)
Infection:
  Algerita
  Chaparro Amargosa
  Echinacea
  Elephant Tree
  Hollyhock
  Mesquite
  Yerba Mansa
  Intestinal Disinfectant (#10)
  Diarrhea Formula (#13)
(*See* Anti-
microbial/antiseptic.)
Irritable Bowel Syndrome
(IBS):
  Canadian Fleabane
  Silk Tassel
  Wild Lettuce
Tonic:
  Agave
  Jojoba
  Yerba Mansa
(*See* Convalescent Aid,
Tonic.)

Ulcers:
  Hollyhock
  Jojoba
  Mesquite
  Sangre de Drago
  Shepherd's Purse
  Sumach
  Yerba Mansa
  (*See* Hemostatic, Stomach.)

INTOXICANTS
  Syrian Rue
  Wild Lettuce

KIDNEYS
Nephritis, Pyelitis:
  Hollyhock
  Shepherd's Purse
Stones, Uric Acid:
  Shepherd's Purse

LARYNGITIS
  Acacia
  Hollyhock
  Mesquite
  Sage
(*See* Cough, Throat
Inflammation.)

LAXATIVE/CATHARTIC
General Constipation:
  Agave
  Desert Senna
  Prickly Poppy
  Sangre de Drago
with Pelvic Congestion:
  Ocotillo
with Dry Skin:
  Vervain
  Silver Sage
with Dyspepsia, coated
tongue, poor fat digestion:
  Algerita
  Indian Root
  Lipid Normalizer (#2)

LINAMENT HERBS
  Matarique
  Camphor Weed

172

Pineapple Weed
Sagebrush
Silver Sage
Stillingia
Turkey Mullein

LIVER
Bile and Gall Bladder
Stimulant:
  California Mugwort
  Prodigiosa
  Sagebrush
Gall Bladder Pain:
  Indian Root
  Prodigiosa
  Silk Tassel
Hepatitis/Cirrhosis:
  California Mugwort
  Chaparral
  Milk Thistle
Stimulant/Lipid Metabolism:
  Chaparral
  Milk Thistle
  Lipid Normalizer (#2)
Stimulant/Protein
Metabolism:
  Algerita
  Indian Root
  Milk Thistle
Stimulant/Immunologic:
  Echinacea
  Indian Root
  Stillingia
  Tonic/Protectant:
    Milk Thistle

LYMPH SYSTEM
General:
  Red Root
  Stillingia
  Alterative Syrup (#15)
Lymphadenitis, acute:
  Red Root with Echinacea
Lymphadenitis, chronic:
  Hollyhock
  Red Root
  Stillingia

Congestion in pelvis with
hemorrhoids, vericosities:
  Ocotillo
(*See* Immunologic.)

MAGIC-TALISMAN-RITUAL
  Sagebrush
  Syrian Rue
  Juniper

MERIDIAN DIAGNOSIS
  Red Root
  Vervain

METABOLIC DISORDERS
General:
  Spikenard
Fats and Lipids:
  California Mugwort
  Chaparral
  Milk Thistle
  Ocotillo
  Prodigiosa
  Puncture Vine
  Lipid Normalizer (#2)
Proteins:
  Algerita
  Indian Root
  Matarique
  Prodigiosa
Immunologic:
  Echinacea
  Elephant Tree
(*See* Liver, Stress.)

MOUTH/GUM INFLAMMATIONS
  Acacia
  Añil del Muerto
  Cañaigre
  Echinacea
  Encelia
  Elephant Tree
  Prickly Pear
  Ratany

Sangre de Drago
Sumach
Yerba Mansa
Mouth Wash (#12)

MOXIBUSTION
California Mugwort

MUSCLE/JOINT PAIN
External:
Camphor Weed
Chickweed
Escoba de la Vibora
Matarique
Turkey Mullein
Yerba Mansa
Emollient Poultice (#9)
Internal:
Echinacea
Escoba de la Vibora
Yerba Mansa
Yucca
Fatigue, Exhaustion:
Ocotillo
Turkey Mullein
Yerba de Alonso Garcia
(*See* Arthritis, Bath/Sauna,
Linaments.)

NAUSEA
Desert Lavender
Pineapple Weed
Spikenard
Yerba Mansa

NEURALGIA
Passion Flower
Prickly Poppy
Vervain
(*See* Sedative/Tranquilizer.)

NOSE
Bleeding:
Periwinkle
Red Root
Shepherd's Purse
Nasal Spray:
Sumach

Echinacea
Yerba Mansa

PANCREAS
Algerita
Indian Root
Prodigiosa

PARASITES
Amebiasis, Giardiasis:
Chaparro Amargosa
Intestinal Disinfectant (#10)

PORTAL CONGESTION
Milk Thistle
Ocotillo

POULTICE
Anti-inflammatory:
California Mugwort
Chickweed
Hollyhock
Mallow
Ocotillo
Sage
Emollient Poultice (#9)
Drawing:
Chimaja
Hollyhock
Mallow
Prickly Pear
Piñon

PREGNANCY/
POSTPARTUM
Morning Sickness:
Acacia
Mesquite
Passion Flower
Pregnancy Back Pain:
Buckwheat Bush
Prepartum Tonic:
Spikenard
Birthing:
Cotton Root
Vervain
Postpartum Bleeding:
Buckwheat Bush

174

Cotton Root
Shepherd's Purse
Vervain
Postpartum recuperation,
Uterine Stimulation:
  Cotton Root
  Mallow
Postpartum Sitz Bath:
  Manzanita
  Sangre de Drago
  Yerba Mansa
Lochia Stimulation:
  Spikenard
Nursing Stimulation:
  Cotton Root
  Vervain
Nursing Sores:
  Sumach
Mastitis:
  Cotton Root
  Mastitis Formula (#6)
  Emollient Poultice (#9)
Weaning:
  Sage
Newborn Wash:
  Mallow
  Yerba Mansa
Colic:
  Desert Lavender
  Passion Flower
  Pineapple Weed
  Vervain
Diaper Rash:
  Sagebrush
  Yerba Mansa
(*See* Reproductive System—
Female.)

P.M.S. (Premenstrual Syndrome)

Water retention, protein/fat craving:
  Chaparral
  Yerba Mansa
  Lipid Normalizer (#2)
Irritable, gloomy, low "chi":
  Anemone

Intestinal, lower back, uterine aches:
  Silk Tassel
(*See* Reproductive System—
Female.)

PSYCHIATRIC/ EMOTIONAL DISTRESS

Insomnia (with nervousness, morbidity, wan/weak physically):
  Anemone
Mood elevator in physical depression:
  Syrian Rue
Depression, nervousness in strong individuals:
  Wild Oats

REPRODUCTIVE SYSTEM—FEMALE

Antimenstrual:
  Buckwheat Bush
  Desert Lavender
  Escoba de la Vibora
  Periwinkle
  Red Root
  Shepherd's Purse
Bartholin Gland Cysts:
  Yerba Mansa
Cramps:
  Marsh Fleabane
  Passion Flower
  Pineapple Weed
  Silk Tassel
Fibrocystic Breast Disease:
  Cotton Root with Red Root
Menopause, Palpitations:
  Prickly Poppy
Menopause, Hot Flashes, Sweating:
  Canadian Fleabane
Menorragia/metrorragia:
  Shepherd's Purse
Ovarian Cysts:
  Red Root

Promenstrual:
Anemone
Chimaja
Marsh Fleabane
Spikenard
Uterine Stimulant:
Cotton Root
Vaginitis/cervicitis:
Hollyhock
Echinacea
Jojoba
Manzanita
Ocotillo
Sumach
Yerba Mansa
(*See* P.M.S.)

REPRODUCTIVE
SYSTEM—MALE
Benign Prostatitis:
Cotton Root
Hollyhock
Ocotillo
Prickly Poppy
Hydroceles:
Red Root

RESPIRATORY
Asthma (with Chronic
Inflammation):
Jojoba
Piñon
Prickly Pear
Spikenard
Turkey Mullein (external)
Asthma (with Rapid
Breathing, Dry Chest):
Night Blooming Cereus
Passion Flower
Bronchitis (Early Stages, dry
membranes, hectic
breathing):
Night Blooming Cereus
with Wild Cherry
Passion Flower
Spikenard
Yerba Mansa

Immunostimulant (#8)
Bronchitis (Waning Stages,
Touch Mucus):
Condalia
Encelia
Elephant Tree
Piñon
Emphysema/Bronchiectasis:
Jojoba
Prickly Pear
Spikenard
Turkey Mullein (external)
Pleurisy:
Stillingia
Alterative Syrup (#15)
(*See* Colds/Flu, Cough,
Expectorant, Laryngitis,
Sinuses, Throat
Inflammations, Tonsillitis.)

SALIVA
Stimulant:
Algerita
Echinacea
Indian Root
Stillingia
Vervain

SALVE/OINTMENT
HERBS
Añil del Muerto
Camphor Weed
Chaparral
Chickweed
Cypress
Desert Willow
Echinacea
Encelia
Epazote
Elephant Tree
Eucalyptus
Juniper
Pineapple Weed
Piñon
Prickly Poppy
Ratany
Sage

Sagebrush
Sange de Drago
Sumach
Yerba Mansa
(*See* Counterirritant,
Demulcent/Emollient,
Poultice Herbs.)

SCIATICA
Prickly Poppy

SEDATIVE/
TRANQUILIZER
Insomnia (with nervousness,
gloom):
  Anemone
Insomnia (with
cardiopulmonary
excitement):
  Passion Flower
  Vervain
Insomnia and fever, just
getting sick:
  Vervain
  Wild Lettuce
Insomnia in Children:
  Passion Flower
  Pineapple Weed
  Vervain
Analgesic/Muscle Relaxant:
  Prickly Poppy
  Silk Tassel
  Turkey Mullein
  Wild Lettuce
(*See* Antispasmodic.)

SINUSES
Suppress:
  Canadian Fleabane
  Sage
  Sumach
  Yerba Mansa
Stimulate:
  California Mugwort
  Eucalyptus
(*See* Allergies, Eyes,
Respiratory System,Colds/Flu.)

SKIN
Abrasions:
  Condalia
  Chaparral
  Cypress
  Desert Willow
  Echinacea
  Eucalyptus
  Ratany
  Sage
  Sagebrush
  Sangre de Drago
  Spikenard
  Yerba Mansa
  Mouthwash (#12)

Anti-inflammatory:
  Cañaigre Leaves
  Echinacea
  Mesquite
  Pineapple Weed
  Prickly Pear
  Prickly Poppy
  Sage
  Emollient Poultice (#9)

Antimicrobial:
  Algerita
  Anemone
  Camphor Weed
  Chaparral
  Cypress
  Desert Willow
  Epazote
  Eucalyptus
  Piñon
  Sage
  Sagebrush
  Spikenard
  Trumpet Creeper
  Yerba Mansa

Boils, Abcesses:
  Anemone (external)
  Echinacea
  Hollyhock
  Piñon

177

Sagebrush
Stillingia
Emollient Poultice (#9)
Burns, Sunburns:
  Cañaigre
  Condalia
  Cypress (roasted)
  Prickly Pear
  Prickly Poppy
  Sangre de Drago
  Sunburn Treatment (#4)
Splinters:
  Piñon
Tineas/Fungus infections,
external:
  Cypress
  Desert Willow
  Epazote
  Trumpet Creeper
  Yerba Mansa
Tineas/Fungus Infections,
internal:
  Stillingia
Ulcerations:
  Cañaigre
  Chaparral
  Cypress (roasted)
  Echinacea
  Hollyhock
  Piñon
  Prickly Pear
  Spikenard
  Sumach
  Yerba Mansa
Psoriasis, Exfoliative
Dermatitis:
  Syrian Rue
Warts:
  Prickly Poppy

SMOKING HERBS
  Desert Willow
  Manzanita
  Prickly Poppy
  Sagebrush

SOAP/SHAMPOO
Agave
Condalia
Soapberry
Yucca

SPICE
Chimaja
Epazote
Juniper
Sage

SPLEEN
General:
  Milk Thistle
  Red Root
Splenomegaly, Mild:
  Red Root

STIMULANT
  Marsh Fleabane
  Indian Root
  Red Root Leaves
  Syrian Rue

STOMACH
Bitter Tonic:
  Agave
  Algerita
  Indian Root
  Prodigiosa
  Vervain
Cramps:
  Chimaja
  Desert Lavender
  Passion Flower
  Pineapple Weed
  Silk Tassel
Gas/Fermentation:
  Agave
  Echinacea
  Eucalyptus
  Marsh Fleabane
  Passion Flower
  Pineapple Weed
  Yerba Mansa

Indigestion/Dyspepsia:
  Acacia
  Agave
  Chimaja
  Desert Lavender
  Escoba de la Vibora
  Eucalyptus
  Mallow
  Mesquite
  Juniper
  Silver Sage
Hyposecretions:
  Algerita
  Indian Root
  Matarique
  Juniper
  Prodigiosa
  Vervain
Hypersecretions:
  Desert Lavender
  Hollyhock
  Mesquite
  Silver Sage
Ulcers:
  Desert Lavender
  Hollyhock
  Mesquite
  Sangre de Drago
  Shepherd's Purse (bleeding)
  Sumach
  (*See* Intestinal Tract,
  Nausea.)

STRESS
General:
  Spikenard
Adrenergic:
  Indian Root
  Wild Oats
Adrenergic, with
Palpitations:
  Turkey Mullein

TANNING
  Cañaigre
  Sangre de Drago

TEA (FOR THE TASTE)
  Cenizo
  Cliff Rose
  Jojoba "Coffee"
  Ocotillo Flowers
  Red Root Leaves
  Piñon
  Sumach Berries
  Yerba de Alonso Garcia

TENDONITIS
  Echinacea

THROAT INFLAMMATION
Acute:
  Acacia
  Buckwheat Bush
  Cañaigre
  Echinacea
  Hollyhock
  Eucalyptus
  Mallow
  Mesquite
  Ratany
  Sage
  Sangre de Drago
Chronic:
  Jojoba
  Piñon
  Red Root
  Stillingia
  Yerba Mansa
(*See* Cough, Laryngitis,
Respiratory System,
Tonsillitis.)

TONIC
Gastrointestinal:
  Agave
  Algerita
  Yerba Mansa
Immunologic:
  Echinacea
  Elephant Tree
Liver:
  Algerita

Milk Thistle
Gall Bladder:
  Prodigiosa
(*See* Convalescent Aid,
Metabolic, Stress.)

TONSILLITIS
  Mallow
  Ratany
  Red Root

URINARY TRACT
Cystitis/Urethritis, General:
  Buckwheat Bush
  Cypress
  Elephant Tree
  Hollyhock
  Mallow
  Prickly Pear
Cystitis/Urethritis, Chronic:
  Jojoba
  Mallow
  Ocotillo
  Yerba Mansa
Cystitis/Urethritis, with
Frequent Pain:
  Prickly Poppy
  Silk Tassel
  Yerba Mansa
Cystitis/Urethritis,
Antimicrobial:
  Chimaja
  Juniper
  Manzanita,
  Urinary Disinfectant (#7)
Cystitis/Urethritis with
Hyperacidity:
  Chimaja
  Juniper
  Shepherd's Purse
Cystitis/Urethritis with
Hyperalkalinity:
  Manzanita
Cystitis/Urethritis, from
catheterization:
  Mallow
  Manzanita

Hematuria:
  Canadian Fleabane
  (chronic)
  Shepherd's Purse (acute)
Phosphaturia:
  Shepherd's Purse
(*See* Kidneys.)

VERICOSE VEINS
  Milk Thistle
  Ocotillo
  Vervain
(*See* Cardiovascular, Liver,
Lymph System.)

VERMIFUGE
Roundworms:
  Epazote
(*See* Intestinal tract,
Parasites.)

VETERINARY
Horses:
  Acacia
  Condalia
  Echinacea (tendonitis)
  Shepherd's Purse
  (hematuria)
Cattle:
  Condalia
General birthing:
  Cotton Root
  Vervain
Weaning:
  Sage

WOOD CARVING
  Manzanita

PLANT INDEX

CREDITS

The following color plates are photographed by Rodney G. Engard, courtesy of the Tucson Botanical Gardens:

Acacia, Anemone, Chaparro Amargosa, Condalia, Hollyhock, Puncture Vine, Sangre de Drago, and Yerba Mansa

The following color plates are courtesy of the University of Arizona:

Agave, Añil del Muerto, Buckwheat Bush, Cañaigre, Cenizo, Chaparral, Desert Lavender, Desert Senna, Desert Willow, Elephant Tree, Jojoba, Milk Thistle, Ocotillo, Passion Flower, Prickly Pear, Ratany, Sumach, Tronadora, and Yucca.

ABOUT THE AUTHOR

Michael Moore has been a practicing herbalist since 1968, as a merchant, picker, therapist, teacher, and writer. His books include *Medicinal Plants of the Mountain West* (Museum of New Mexico Press), *Los Remedios de la Gente,* and *Herbal Repertory in Clinical Practice.* Besides directing the School of Clinical Herbology in Santa Fe, New Mexico (1980–1985), he was also a primary instructor in both herbology and physiology for the Santa Fe College of Natural Medicine, the Institute of Traditional Medicine, and the Southwest College of Acupuncture. He has been on the teaching staff of the California School of Herbal Studies, the Platonic Academy, and Humber College, Ontario. He has been a lecturer for the University of New Mexico School of Medicine, University of Arizona School of Medicine, National College of Naturopathic Medicine, Palmer College of Chiropractic, New Mexico Pharmaceutical Association, the College of Santa Fe, the New Mexico Acupuncture Association, and the Arizona Naturopathic Medical Association. For fourteen years he was the owner of Herbs, Etc. of Santa Fe and is presently co-owner, with partner Susan Mullen, of Bisbee Botanicals in Gila, New Mexico.